Giovanni Battista Ercolani

The Utricular Glands of the Uterus,

and the glandular organ of new formation which is developed during pregnancy in

the uterus of the Mammalia, including the human species

Giovanni Battista Ercolani

The Utricular Glands of the Uterus,
and the glandular organ of new formation which is developed during pregnancy in the uterus of the Mammalia, including the human species

ISBN/EAN: 9783337373382

Printed in Europe, USA, Canada, Australia, Japan

Cover: Foto ©berggeist007 / pixelio.de

More available books at **www.hansebooks.com**

THE
UTRICULAR GLANDS OF THE UTERUS,

AND

THE GLANDULAR ORGAN OF NEW FORMATION WHICH
IS DEVELOPED DURING PREGNANCY IN THE
UTERUS OF THE MAMMALIA, INCLUD-
ING THE HUMAN SPECIES.

BY

PROF. GIOVANNI BATTISTA ERCOLANI,

PERMANENT SECRETARY OF THE ACADEMY OF SCIENCES OF BOLOGNA; CORRESPONDING MEMBER
OF THE ACADEMY OF MEDICINE OF PARIS AND BRUSSELS, AND OF THE
GYNECOLOGICAL SOCIETY OF BOSTON, ETC.

TO WHICH IS APPENDED

HIS MONOGRAPH UPON THE UNITY OF THE ANATOMICAL TYPE OF
THE PLACENTA IN ALL THE MAMMALIA, AND THE PHYSIO-
LOGICAL UNITY OF THE NUTRITION OF THE FŒTUS
IN ALL THE VERTEBRATES.

ALSO,

A GENERAL SUMMARY AND CLASSIFICATION,

WRITTEN EXPRESSLY FOR THIS EDITION.

WITH A QUARTO ATLAS OF FIFTEEN PLATES,

ENGRAVED BY BETTINI, AND REPRODUCED BY
THE HELIOTYPE PROCESS.

TRANSLATED FROM THE ITALIAN UNDER THE DIRECTION OF

HENRY O. MARCY, A. M., M. D.,

VICE-PRESIDENT OF THE AMERICAN MEDICAL ASSOCIATION; MEMBER OF THE MASSACHUSETTS
MEDICAL SOCIETY; FELLOW OF THE AMERICAN ACADEMY OF MEDICINE,
BOSTON GYNECOLOGICAL SOCIETY, ETC.

BOSTON:
HOUGHTON, OSGOOD AND COMPANY.
The Riverside Press, Cambridge.
1880.

Copyright, 1880,
By H. O. MARCY, M. D.

RIVERSIDE, CAMBRIDGE:
STEREOTYPED AND PRINTED BY
H. O. HOUGHTON AND COMPANY.

TRANSLATOR'S PREFACE.

The translation and editing of the several monographs of Prof. G. B. Ercolani has been undertaken from the conviction that their publication in English would be of service to science.

The opinions, at a greater or less length, of this distinguished scientist are quoted in nearly all the modern text-books treating upon this subject, and yet so indefinitely in many instances, and even erroneously, that it is evident authors have not familiarized themselves with the elaborate and painstaking studies of Professor Ercolani.

So important are these demonstrations and the truths derived therefrom, so radically different are his teachings from the time-honored views still held and generally taught, that I have felt the medical profession and students of natural history would gladly avail themselves of the opportunity of carefully examining these original investigations. Especially have they seemed to me valuable because of the attention which the study of the human placenta has received of late by many careful observers, and still more so since their conclusions are by no means unanimous. Anatomists and physiologists have long experienced the great necessity of a clearer knowledge upon many of the reproductive processes. More recently the gynæcologist seeks in the changes of the utero-gestation period a solution to many of the pathological questions demanding his attention.

Since the classic work of William Hunter a century has passed with only a very few additions to what was actually known in reference to the nutrition of the embryo.

Professor Ercolani's monographs appeared originally in the Transactions of the Academy of Science of Bologna. The memoir was translated into French by Prof. R. Bruch and Dr. R. Andreini, for whom the Appendix here given was written, and has received an award from the Academy of Sciences at Paris. I am under a personal obligation to these gentlemen, since it was the study of their translation which prompted its reproduction for the English reader. To my friends, M. Smead and S. S. Jacobs, who have so very thoroughly performed the work of translators, I would express my gratitude, for without their faithful aid this translation would never have been presented to the public. Whatever faults of diction or liberty taken with the translation for clearness of expression is assumed by myself. The work on my own part has been a labor of love. Stimulated by the teachings of the learned Professor Martin, of Berlin, during the last decade, I have improved every opportunity for original investigations in this direction, and have verified a number of the observations made by Ercolani.

Original study never lacks interest or loses value. The author declares that "Science is pure and simple truth," and states that his aim has been to establish the fact that the maternal portion of the placenta of mammals and the human species is always a glandular organ of new formation, and is developed for the secretion of a fluid, which serves for the nutrition of the fœtus.

The placenta consists in all animals of two distinct fundamental parts, the fœtal portion, vascular and absorbent, the maternal, glandular and secretory. In no case do the vessels of the mother anastomose with the vessels of the fœtus. He elucidates the nature and metamorphosis of the utricular glands of the uterus, and resolves affirmatively the question as to the presence of a uterine mucous membrane in woman.

In the same manner as, in the first period of extrauterine life, the young is nourished by the mother's milk, absorbed by the villi of the intestine, so, during intra-uterine life, the fœtus finds its nutriment in the maternal fluid secreted by the glandular organ. In the eggs of birds and reptiles there is elaborated beforehand, by the mother, and stored up, all the nutritive material necessary for the development of embryonic life.

New teachings are rarely announced without finding opponents, and this is true of the work presented. Professor Ercolani writes, "This opposition has been advantageous to me; I was so sure of the facts that I should have perhaps remained inactive and not have taken the trouble to observe and follow the different periods of development of the glandular organ in animals and in woman. I brought forward these new deductions from my recent observations in the public conference where I replied to the objections of Professors Albini, Palladine, and Ohel." In this discussion he stated in the frankest and most conscientious manner certain errors of interpretation in the first facts announced.

By these researches is opened an almost unexplored field in the pathology of gestation, and this, too, is essentially the medical and more practical side of the subject. In the abnormal development of the placenta and in the modifications of the nutrition of the fœtus will be found causes hitherto unknown of embryonic disease and arrest of intra-uterine life.

Therefore it will be seen that the work of the Bologna professor has an intimate connection and bearing upon anatomy and physiology, chemistry and pathology, embryology and anthropology, biology and obstetrics. It destroys ancient and classic errors; it demonstrates an important new anatomical fact; it teaches a new physiological function, and clearly shows a simple and fundamental plan of embryonic life.

The evident impartiality of the author, as shown in his numerous observations, the multiplicity of facts produced, the modest and conscientious expression of opinion, and the beautiful illustrations of his studies in the accompanying volume of plates appeal to the unprejudiced reader, and carry conviction that the deductions presented are the results of thoughtful labor, and not preconceived theories which he has endeavored to demonstrate. The establishment of such facts will cause the name of Ercolani to be classed with the great benefactors of science and be handed down to coming generations, honored alike with Eustachius, Malpighi, Morgagni, and other distinguished anatomists of the early Italian school.

HENRY O. MARCY.

CAMBRIDGE, *January*, 1880.

CONTENTS.

CHAPTER I.

                                         PAGE.

INTRODUCTION 1–5

CHAPTER II.

THE UTRICULAR GLANDS OF THE UTERUS IN THE MAMMALIA IN THE GRAVID AND NON-GRAVID ANIMAL.

Their development and function 6–26

CHAPTER III.

ON THE GLANDULAR ORGAN, OR THE MATERNAL PLACENTA, IN ANIMALS WITH A VILLOUS OR DIFFUSED PLACENTA.

Opinions of the ancients. — Histological changes of the uterine mucous membrane. — Illustrations drawn from the mare, as seen in Plate III., Figs. 1, 2. — Demonstrations, from the studies of comparative anatomy, that there exists a true uterine mucous membrane in all animals, including the human species. — The transformation which the uterine mucous membrane undergoes during pregnancy to give rise to the neo-formation of a transitory glandular organ constituting the maternal part of the placenta. — Proliferation of sub-mucous connective tissue. — Glandular follicles of new formation and their secretory function 27–46

CHAPTER IV.

THE GLANDULAR ORGAN, OR MATERNAL PLACENTA, IN ANIMALS WITH A MULTIPLE PLACENTA, AS IN THE RUMINANTS, OR THE UTERINE COTYLEDONS IN THOSE ANIMALS.

Opinions held by the earlier writers. — The fluid furnished by the cotyledons: its character, chemical composition, etc. — Fabricius and his followers taught the direct communication of the maternal vessels with those of the fœtus in the cotyledons. — Harvey maintained that the fluid secreted by the cotyledons was absorbed by the villi of the fœtal placenta. — Formation and development of the cotyledons. — The structure and function of the utricular glands. — Histological characters as illustrated by Plates V. and VI. 47–64

CHAPTER V.

THE GLANDULAR ORGAN, OR MATERNAL PORTION OF THE PLACENTA, IN ANIMALS WITH A SINGLE PLACENTA.

Is the placenta entirely fœtal, or is it distinguishable into two parts, one maternal and the other fœtal? — Harvey, Fabricius, Wharton, Needham, Malpighi, of the early writers. — Views of Von Baer, Sharpey, Weber, Bischoff, Eschricht, Panizza, and others. — The doctrine erroneous that the exchange of materials for the nutrition of the fœtus is carried on by the process of endosmose and exosmose through the walls of the maternal and fœtal vessels. — Two different periods of fœtal nutrition. — A glandular organ of new formation demonstrated by the aid of studies upon animals having a diffused or villous placenta. — The placenta of animals, although single, as distinct from that of the human species. — Plate VII. shows sections from the placenta of the rabbit, teaching its histological development. — Among the carnivora the dog has been selected, and Plates VIII. and IX. teach that the placenta is formed in the same manner as in the rabbit. 65–101

CHAPTER VI.

THE HUMAN PLACENTA.

The changes which the surface of the uterus undergoes at the placental site. — Opinions held as to the mode of formation and structure of the uterine decidua. — The decidua serotina so peculiar as to prevent its being confounded with the uterine and reflected decidua. — The exterior membrane surrounding the villi of the chorion is furnished by the decidua serotina. — Hypertrophy and hyperplasia of the anatomical elements constituting the connective tissue, and the transformation which gives rise to the development of special neutral cells, from which is formed the new glandular organ, or maternal portion of the placenta. — Plate X. represents a vertical section of the uterine surface of the placenta. — Distinctions between the human placenta and that of animals. — The utero-placental vessels and venous sinuses. — Lacunar circulation. — Ectasic condition of the maternal vessels. — In all the mammalia, including the human species, the maternal portion of the placenta is always a glandular organ of new formation 102–129

CHAPTER VII.

I.

CONCLUSIONS RELATIVE TO THE UTRICULAR GLANDS AND THE MUCOUS MEMBRANE OF THE UTERUS.

Erroneous views of anatomists and physiologists who have held that in certain animals there were two species of uterine glands. — The uterine decidua of woman a product of the utricular glands. — The glandular secretion for the nutrition of the embryo until the development of the new glandular organ which constitutes the maternal portion of the placenta 130–136

II.

CONCLUSIONS UPON THE GLANDULAR ORGAN OF NEO-FORMATION, OR MATERNAL PORTION OF THE PLACENTA, IN THE MAMMALIA AND IN THE HUMAN SPECIES.

The placenta composed of two parts entirely distinct both in structure and in function : the fœtal portion, vascular and absorbent; the maternal portion, glandular and secretory. — In no case do the vessels of the mother come in contact with those of the fœtus. — The fœtus always nourished by the fluid secreted by the glandular organ and absorbed by the villi of the chorion 137–145

APPENDIX.

I.

THE FORMATION OF THE MATERNAL OR GLANDULAR PORTION OF THE PLACENTA IN THE HUMAN SPECIES AND IN CERTAIN ANIMALS 147–150

II.

FORMATION OF THE MATERNAL PLACENTA IN THE COW, SHEEP, MOLE, AND HIND 150–156

III.

FORMATION OF THE SINGLE PLACENTA IN THE CAT, THE HARE, AND THE GUINEA-PIG 157–168

IV.

FORMATION OF THE PLACENTA IN WOMAN AND IN THE MONKEY 169–174

MONOGRAPH.

THE UNITY OF THE ANATOMICAL TYPE OF THE PLACENTA IN THE MAMMALIA AND IN THE HUMAN SPECIES, AND THE PHYSIOLOGICAL UNITY OF THE NUTRITION OF THE FŒTUS IN ALL THE VERTEBRATES.

I.

Comparative anatomy reveals the structure of the human placenta and demonstrates the unity of the anatomical type in all animals. — Newly observed facts explain the origin of the elements of the decidua and the maternal portion of the placenta, hitherto lacking positive demonstration. — The complete destruction of the uterine mucous membrane, and the subjacent parts, indispensable in all cases to the establishment and development of the neo-formative changes, from which will result the maternal portion of the placenta. — Plate I. represents the gravid uterus of the rabbit about fifteen days after conception, at the beginning of the formation of the placenta. — Plates II., III., and IV. further illustrate the destructive and neo-formative changes in different animals. — Deductions and conclusions 175–218

II.

PHYSIOLOGICAL UNITY CONTROLLING THE NUTRITION OF THE FŒTUS IN ALL VERTEBRATES.

Maternal aliment, and the means by which it may be appropriated by the embryo. — Represented by the yolk of the egg in the oviparous animals and by the placenta in mammals. — Nature accomplishes her purpose for the nutrition of the embryo by materials furnished by the mother. — The Marsupialia. — Plate V. illustrates in diagram the unity of type in the various forms of placenta. — Relation between the fœtal and maternal parts by simple proximity, contact, or by intimate cohesion. — Relations between the fœtal and the maternal portion of the placenta in woman identical with those observed in single placentæ in the other mammifera. — Summing up of principal facts. — One law, a physiological modality governs the nutrition of the fœtus in all the vertebrates 218–271

SUMMARY AND CLASSIFICATION.

Acotyledonous placentæ: single, villous, and diffused. — Complicated villous. — Localized villous. — Cotyledonous placentæ of incomplete and with complete vascularization. — With ectasia in the vessels of the maternal portion of the placenta. — Table of classification of the vertebrates. — Conclusions 272–296

UTRICULAR GLANDS OF THE UTERUS.

CHAPTER I.

INTRODUCTION.

The observations which I have the honor to present upon the formation and structure of the placenta in the human species and among the mammalia have led me to such different conclusions from those which are generally admitted at the present day by anatomists and physiologists that I feel it necessary, at the outset, to appeal to the benevolent attention of my readers. I shall further need indulgence, because I shall be obliged to describe very minutely a great number of researches which concur, in their whole, to explain the general idea of my work. For this reason, then, and for the sake of suitable method also, it seems advisable to begin with illustrating this general idea by a few diagrams, and afterwards to discuss the researches and observations of which that idea is the result, and which I will sum up as follows : There is produced in the pregnant uterus of mammals, including the human species, a glandular organ of new formation. This organ constitutes one of the two fundamental portions of the placenta, that is to say, the maternal portion, with which the foetus is brought into intimate relation by the villi of the chorion, which composes the other portion of it, or the fœtal part.

The villi of this latter part of the placenta penetrate always and obviously into the glandular organ or maternal part, in order to absorb the fluid which is there secreted, and thus to furnish the fœtus with the materials necessary for its nutrition.

The typical form of the new glandular secreting organ does not depart from the common form of a simple glandular follicle of the animal organism. Likewise the typical form of the fœtal placenta, or absorbing portion, is that of a vascular loop, more or less elongated, or of a villus. This is observed, with a few rare exceptions, in the three fundamental species of placenta admitted by anatomists under the names of disseminated, diffused, or villous placenta, multiple placenta, and single placenta.

In the diagrams, I have represented vertical sections of the uterus and of the placenta of animals and of the human species, in order that the relations always existing between the two portions of the placenta may be more clearly demonstrated.

Figure 1, Plate I., is taken from the mare, as an example of the villous placenta. At the top is the chorion (a), from which issue the vascular pencils or tufts and the villi (b, c, c) of the fœtal portion, which penetrate into as many simple glandular follicles (d) of the maternal portion, developed, during pregnancy, over the whole surface of the uterus, the walls of which are indicated by $e\ e$. This is the simplest possible type of the double structure of the placenta.

In Figure 2 of Plate I. I have in like manner indicated the same portions of the multiple placenta, taking as the type that of the cow, which offers the simplest form of this kind of placenta common to ruminants. Although the glandular organ is here modified, it does not lose its elementary form of simple follicle. The only change is in the degree of proximity and the position of the follicles.

In the disseminated or diffused placenta we have seen the follicles placed vertically over the whole internal surface of the uterus; here, on the contrary, they are parallel to the same surface, and superimposed over each other, at the points where the various placentæ are developed (d, d). The relations of the villi with the follicles are the same as in the preceding case.

As for the single placenta, we must not confound, in one type, that of certain animals and that of the human species.

In the dog and cat (Plate I., Fig. 3) the typical form of the glandular follicle is not lost; but, instead of being repeated in its simple form, as we have seen it in the cow, it is extraordinarily elongated into tubular glands, as it were, which are packed closely, by their walls, against the villi of the fœtal placenta. The opening of the follicles on the surface of the placenta occurs at the place where the villi of the chorion penetrate (g). Their extremity, or closed end, is visible in the interior of the placenta towards its uterine face (g, g). Nevertheless, it is impossible to follow or isolate an entire follicle from its orifice to its termination, on account of the complicated and sinuous structure of the enteriform loops, of their very intimate proximity to each other, and of their numerous connections with the substance of the placenta.

It is in the human species that the structure of the glandular organ, or maternal placenta, differs in the most remarkable manner from the typical form of the simple glandular follicle. When discussing this organ, I shall point out the great differences which distinguish it from that of the lower orders. For the present, I will only state that in the human maternal placenta the fundamental parts of the organ, that is to say the walls and the cells, — in a word, the secreting organ and the secretion itself, — are persistent; but all that relates to the form of a glandular follicle is completely lost.

Figure 4 represents the human placenta. Contrary to what we observe in animals, that portion of the uterine surface which is in contact with the placenta is covered with a peculiar membrane of new formation, known to anatomists under the name of decidua serotina (f). This membrane, produced by the proliferation of the cells of the superficial or submucous connective tissue of the uterus, is the stroma whence originates the glandular organ which supports and envelops the villi of the foetal placenta in all their numerous subdivisions (d, d).

The glandular organ accompanies the villosities as far as the chorion; having reached this point, it loses its glandular structure and becomes altogether fibrous, in order to attach firmly to the chorion itself the vessels of which the umbilical cord is formed (g).

By this prefatory general statement of the structure of the placenta in mammals and the human species, I have marked out the order which I shall follow in my work. As is well known, it has been taught and admitted by many authors that the uterine glands of animals, at least, if not those of woman, play a very important part in the formation of the placenta. It will therefore be fitting to treat first of these glands, then explain the structure of the placenta, — whether it is disseminated or villous, as in the solipeds, or multiple, as in the ruminants, or single, as in the carnivora and rodents, — and conclude with the human placenta.

I shall, moreover, preface each of these divisions by a few historical researches, for the special purpose of elucidating such of the new observations as appear to me of most importance.

CHAPTER II.

THE UTRICULAR GLANDS OF THE UTERUS.

MALPIGHI was the first to make known the existence of the uterine glands. We learn from him [1] that, having returned from Messina to Bologna, he resumed his well-beloved anatomical studies, and began at that time his researches upon the structure of the uterus. In describing these researches to Sponius, in a letter dated from Bologna in 1681,[2] he wrote: —

"Uterus interius membranâ quâdam ambitur, quæ minima et innumera habet orificia, glutinosum, mucosumque fundentia humorem, quo uterus ipse et vagina perpetuò madent. Quare compresso utero prosilit hujusmodi mucosus ichor. Patent autem hæc excretoriorum vasorum ora, si diù interior membrana aquâ maceretur, et in ovibus præcipue obvia sunt; quare probabile est, subvitellina exarata corpora his orificiis in uteri cavitatem hiare; an vero his minimæ appendantur glandulæ, licet sensus non attingat, ratio tamen ex perpetuâ operandi normâ probabiliter eas suadet."

Although from this it is certain that the illustrious anatomist observed the fact in different species of animals, since he declares that we see in the sheep, better than in any other animal, the openings of the

[1] Opera Posthuma. Venice, 1698, p. 46.
[2] Opera omnia. London, 1686.

mucous membrane of the uterus, yet it will be well also to recall his words upon the opening of the uterine glands into the mucous membrane of the uterus in the cow: —

"In prægnantibus vero et præcipue in vaccis, uteri stygmata obvia fiebant." [1]

These few words secure to Malpighi the merit of another quite important observation, namely, that the uterine glands increase in volume during pregnancy.

It is true that it was only from the orifices in the uterus that Malpighi inferred the existence of the glands, of which he had no demonstration by the senses. But we shall not venture to reproach him for this, if we reflect upon the limited means of investigation at his disposal, and remember that the observation was not repeated till one hundred and seventy years later. Besides, the learned anatomists who first repeated it, as the brothers Weber, at first considered the uterine glands as villi of the decidua serotina.[2] It was not until some years later that they corrected their error, recognizing them for what they really are, and denominating them utricular glands of the uterus.

Among the moderns, Baer[3] also, when he observed these same glands in the uterus of the sow and of the cow, took them for lymphatic vessels; and it was not until nine years later, after the observations and rectifications of the brothers Weber, that he recognized

[1] Opera Posthuma, loc. cit.

[2] Disquisitio Anatomica Uteri et Ovariorum Puellæ Septimo a Conceptione Die Defunctæ. Halle, 1830.

[3] Untersuchungen über die Gefässverbindung zwischen Mutter und Frucht in den Saugethiere. Leipzig, 1828.

the glandular structure of the canaliculi which he had at first considered as vessels.[1]

Like Malpighi, Baer had observed the orifices in the mucous membrane of the uterus; but he had not been able to follow the canals to their extremity, for in the sow they are very long. It is evident that the brothers Weber and Baer, even while using the powerful means of observation which modern science has given us, judged by induction in their first researches, just as Malpighi had done. Now, if we compare the inductions and take into account the discouraging poverty of means at the latter's disposal, we can only regard the great anatomist of Bologna with admiration.

Following the development which the observation of Malpighi received in later times, we see that E. Weber, in the publication of the fourth edition of The Human Anatomy of Hildebrand,[2] claimed for himself the honor of having first recognized the existence of the uterine glands, and of having named them utricular glands, giving, besides, a description of them in the uterus of the cow and the doe.

I think we should not deny to Weber the merit of having established, beyond question, the existence of the uterine glands; but if this is just with regard to Weber, equity also requires us to assign to Malpighi the merit which belongs to him.

In 1834, Burkhardt[3] was acquainted with the labors of Malpighi, but he was ignorant of the first observations of Weber and of Baer. So, notwithstanding the

---

[1] Über Entwickelungsgeschichte der Thiere. Königsberg, 1837.
[2] Brunswick, 1832.
[3] Observationes Anatomicæ de Uteri Vaccini Fabrica. Basle, 1834.

teachings of the anatomist of Bologna, he did not recognize the tubular structure of the glands, and he described the uterine glands of the cow under the name of spiral vessels. He repeated his experiments upon the gravid and non-gravid uterus, and confirmed what had already been said by Malpighi, — that they increase in volume during pregnancy; but, not being able, as he acknowledged, to form a precise opinion as to their office, he contented himself with supposing that they were perhaps of some utility to the fœtus.

The later observations of Eschricht[1] confirmed, in a great measure, those of Weber; he added a few upon the uterine glands of dolphins, and he noted some special features in those of cats. His observations upon the latter animal are not exact, and their inaccuracy comes from the fact that Eschricht could not discover the orifice of the glands in the uterus. He does, however, notice the increase of volume which the glands undergo, in that animal, during gestation.

So far, the observations upon the existence of the uterine utricular glands had no other importance than that of the knowledge of an anatomical fact which, though pointed out long ago, had not been demonstrated with the certainty desirable in such matters; and, indeed, apart from this demonstration, nothing had really been added to what Malpighi had said, either as to their existence or as to their increase in volume during pregnancy.

This new anatomical idea did not become the point

---

[1] De Organis quæ Respirationi et Nutritioni Fœtus inserviunt. Copenhagen, 1837.

of departure for a real scientific progress until Sharpey[1] published his researches upon the uterine glands, especially those of the dog. He proved first, that in the dog there exist two species of utricular glands, the simple and the branching. He said that during pregnancy the first do not increase in volume, like the second. He went still further: he taught that the branching uterine glands, which correspond to the point where the ovum is arrested after fecundation, undergo, near their mouths, a partial saccate dilatation, produced by the penetration of a villus of the chorion.

We see that the idea that the fluid secreted by the utricular glands of the uterus served in some manner for the nutrition of the fœtus was confirmed more and more. I say expressly that it was confirmed, for Burkhardt had beforehand suspected it, as it were, and Eschricht had taught more clearly and distinctly that the uterine glands of the pachyderms, ruminants, and cetaceans elaborate a fluid destined for the nutrition of the fœtus.

Bischoff[2] repeated the observations of Sharpey upon the evolution of the embryo in the dog. He gave the name of crypts to the simple uterine glands of Sharpey, and demonstrated the fact of the penetration of the villi of the chorion into the upper parts of the utricular glands. But he added that this fact is observed only during the first periods of gestation, and that, never having verified it later, he could not absolutely affirm it. It remained, however, fully established that, during the first period of intra-uterine

---

[1] The observations of Sharpey are found in a note to the English translation of Müller's Physiology.

[2] Entwickelungsgeschichte des Hundes. Brunswick, 1845.

life, the fœtus is nourished by the liquid secreted by certain glands of the uterus.

E. H. Weber[1] confirmed the researches of Sharpey, describing the partial dilatations of the utricular glands and the attenuation of their walls, destined to favor the contact of the vascular net-work of these glands with that of the fœtal villi, which fill their saccate dilatations. He limited his observations to the gravid uterus of the dog, and declared that he had not been able to demonstrate the same facts in that of woman.

These studies, although lacking the clearness and precision to be desired in such matters, have remained an acquisition to science, and are met with in all recent works on physiology. But before deciding on their exact value, we must await further and more perfect observations.

I ought not, however, to omit the mention of some opinions, published from time to time, which served to emphasize the importance of the uterine glands in the mammifera during gestation.

Thus, the illustrious Professor Gurlt,[2] in speaking of the uterine glands of the mare, affirms that the villi of the chorion enter into openings of these glands, and, for that very reason, they are quite obvious in the gravid uterus.

Colin says[3] that they are the means by which the placental papillæ are united with the uterine mucous

---

[1] Zusätze zur Lehre von Baue und Verrichtungen der Geschlechtsorgane. Leipzig, 1846.

[2] Handbuch der vergleichenden Anatomie der Haussaugethiere. Berlin, 1860.

[3] Traité de Physiologie des Animaux domestiques. Paris, 1856. Vol. ii., p. 561.

membrane, and that this is true of almost all animals; and he quotes in support of this opinion, but not very appropriately, the views of Weber.

To return, however, from assertions to actual observations, Leydig recapitulated a few of those upon the uterine glands.[1] He asserted that the mucous membrane of the uterus is of glandular structure in a great number of mammals. These glands are long and canaliculated in the mare, the sow, and the carnivora, and very long in the ruminants. According to Barkow, they must have a very great development in the seals. Eschricht had already demonstrated their existence in dolphins. Myddleton has found them highly developed in the opossum. Leydig, who had not at first discovered them in the mole, afterwards recognized that they exist in that animal, and that they present a utricular form analogous to the glands of Lieberkuhn. In rats, instead of glands we find very distinct folds of mucous membrane. Yet, he says, considering the subject from another point of view, we may regard the spaces comprised between these folds as colossal glands. We should find the same conditions, in that case, as in the intestinal glands of the batrachians, which consist of alveolar folds of mucous membrane, or of short sacs of considerable size. He affirms that the openings of these glands are infundibuliform in the ruminants; and he repeats what Malpighi had already written, — that during gestation their orifices are so dilated that they can be perceived with the naked eye. Finally, in these later times, Professor Spiegelberg,[2] at the close

[1] Lehrbuch der Histologie des Menschen und der Thiere. Frankfort, 1857.
[2] Zeitschrift für rationelle Medizin von Henle und Pfeufer. Band 21

of some observations made upon the uterus of the ruminants, declares that the openings and the canals of the maternal cotyledons, into which the villi of the chorion penetrate, are nothing but remarkable expansions or dilatations of the uterine glands, and that they serve to establish the intimate union of the mother with the fœtus. But before Spiegelberg, Bischoff had said, in speaking of the allantois of the ruminants, that, during gestation, the numerous elevated points upon the internal surface of the uterus, which are also observed in non-gravid cows, take on a great development, sometimes under the form of depressions, sometimes under that of rounded elevations with numerous excavations; that the latter appear to be formed only by the uterine utricular glands developed into tubes, into which the villi of the chorion are fitted; and that through their walls is carried on the exchange of materials between the blood of the mother and that of the fœtus. This has been the condition of science upon this profound question from Malpighi's time to our own day.

The observations upon the penetration of the villi into the uterine glands have not since been confirmed, and nothing has been added to what Sharpey, Bischoff, and Weber had taught us with regard to but one species of animal with single placenta, namely, the dog. However, the naturalist was almost forced to recognize the importance of such a conclusion, especially after Gurlt had affirmed that this fact was very clear in all cases where the placenta is villous or diffused, as in the mare, and after the statements of Bischoff, and still more after those of Spiegelberg, who taught that the cotyledons of the ruminants were only very remarkable dilatations of the extremity of certain uterine glands.

I regret that, in view of the order and the brevity which it is indispensable to bring into all that remains for me to say, I am forced, for the time being, purely and simply to negative all these observations. I shall point out the facts which have led me to this absolute negation, when I describe the means of union between the mother and the fœtus in the different forms of placenta, and judgment can then be passed upon my researches and my opinion.

For the present, I shall limit myself to a few anatomical considerations upon the utricular glands of the uterus of the different species of animals that I have been able to examine. The type of these glands does not change, in general, in the various species; but it cannot be said that there are no differences to be found. Thus, to begin with the least important, I will remark that if these glands are all formed by a canal of variable length, simply flexuous or tortuous, composed of a thin, fibrous, external membrane, and covered internally by an epithelial layer, nevertheless the thickness of the external membrane and the form of the epithelium vary in quite a remarkable manner. In some animals the glands are formed by a canal that is always uniformly spiral, as in the mare (Plate III., Fig. 1). In the cow, on the contrary, the diameter of the canal, as well as the form of the glands, is variable, owing to the very irregular saccate prolongations, or projecting or goitre-like appendages (Plate II., Fig. 2). In dogs and some other animals the very conspicuous saccate appendages are never absent (Plate II., Fig. 1); and though they are irregular, yet, the projecting or goitre-like form being wanting, their irregularity is less apparent than in the cow, and the term branching, which has been applied to them, is exact (Plate II., Fig. 1, *a*). In the cat the glands are pyriform (Plate IV., Fig. 2), and only upon the largest did I find any expansion or sinuous arrangement at their extremities. Among the *erinaceadæ* of Europe, the utricular glands are formed by a uniform tube, which, after proceeding for a short distance from its orifice in almost a straight line, is rolled up in the shape of a ball, like the sweat glands in man.

In all these animals the utricular glands lie in an oblique direction, and are almost parallel to the internal surface of the uterus. For this reason and on account of their length, it is often very difficult to examine one in its entire course, and to measure its exact extent. The direction of their course is also the cause of the constant obliquity presented by their orifices (Plate III., Fig. 3), which are usually tunnel-shaped (Plate II., Fig. 1, *a*, and Plate IV., Fig 2, *a*). In some other animals the opening is as round as the cavity of the gland, although it presents itself obliquely, as in the mare (Plate III., Fig. 3). In some animals, again, the internal epithelium is paved, as in dogs and cats (Plate II., Fig. 1, and Plate IV., Fig. 2); in others it is cylindrical, as in mares and cows (Plate III., Fig. 3; Plate II., Fig. 2; and Plate V., Fig. 2). Of the utricular glands of the uterus in woman some are simple, others branching; their external membrane is excessively thin, and some anatomists, after M. Velpeau, call it anhistous or formless; the internal epithelium is delicate and cylindrical. It is not as easy to observe them as it is in animals. According to Hirtl,[1] the mucous membrane of the uterine cavity would be represented by these tubular glandules, united by a connective tissue, and by capillary blood-vessels. Yet sections of the glands, parallel or transverse to the mucous membrane of the uterus, have shown me that in the human species they are not so closely pressed against each other as is generally claimed, and that they were drawn from the imagination rather than from nature in the memoir of Weber. This has not, however, prevented Weber's representation from being reproduced in a very great number of recent works upon human and comparative anatomy and physiology.

In all animals, during gestation, the uterine glands increase in volume, and the epithelium becomes more transparent and more delicate.

Anatomists and physiologists have accepted the opinion of Sharpey, confirmed by Weber and Bischoff, that in the cat and dog there exist two species of uterine glands, the simple and the branching, to which some have been inclined to attribute very dissimilar functions: to the simple, that of secreting the uterine mucus; to the branching, that of contributing to the nutrition of the fœtus. It was very interesting to me to verify such an observation; however, as I have before stated, by vertical sections of the uterus only transverse cuts of the glands could be obtained. But it occurred to me that, in order to secure some simple glands, crypts, or follicles, and portions of the branching ones large enough for examination and comparison, it would be necessary, perhaps, to remove, by the aid of a brush and by

[1] Manuale d'Anatomia Topografica. T. ii., p. 112. Milan, 1858.

several washings, the surface layers of the epithelium of the mucous membrane; then to lift up with the forceps the most superficial envelope of the mucous membrane thus denuded, and cut this same envelope at its base, that it might be stretched upon glass. I could then, with the aid of the microscope, easily discover in this thin layer what it was important for me to observe. The experiment succeeded, and it will always be of the greatest value to those who shall desire to examine the uterine glands of animals. By this means, I have several times obtained entire utricular glands of the uterus of the dog as they are represented (Plate II., Fig. 1, *a, a*); but all the experiments that I have repeated by this same method have never enabled me to find a single simple gland or crypt in the uterus of the dog. Since I often obtained complete utricular glands, I ought also to have found, and much more easily indeed, the short glands or crypts, if they had really existed.

While studying in the same manner the uterine mucous membrane of the cat, I thought at first that I recognized the two species of glands of Sharpey and Weber; but on closer examination (Plate IV., Fig. 2, *b, b*) the volume and the variable length of the glands seemed to indicate varying degrees of development or difference of size in the same species of glands rather than to establish two distinct species of glands. Can it be possible that the number of gestations should have sufficient influence to account for the existence or absence in the dog also of what I had observed in the cat? Can it be possible that in certain individuals only the volume of these glands should be variable to such a marked degree? To solve these questions, it would be necessary to devote one's self to a great number of comparative researches, which I have not been able to do. But the observations which I have adduced, although limited, suffice absolutely to refute the assumption of the capital fact of the existence of two distinct species of glands in the uterus of the dog and of the cat,— an alleged fact from which such grave physiological conclusions had been deduced by attributing to each species very different functions. But if we cannot admit two species of uterine glands in the dog and the cat, they are found and easily observed in other species, — for example, in the cow and the sheep. The demonstration of them is positive and easy, by cutting transversely the mucous membrane which covers the rudimentary cotyledons of the non-gravid uterus of those animals; we cannot remove the mucous membrane at the point indicated in the same way as in the dog or the cat. Until our own day, the existence of the crypts or simple glands (which to avoid ambiguity I shall henceforth call simple glandular follicles), had escaped the notice of observers. They are in reality very numerous, and are found agglomerated in the rudimentary cotyledons. Although they are not wanting in the rest of the uterine mucous membrane, they are there fewer and more disseminated.

In vertical and transverse sections made near the internal surface of the cotyledons of the cow (Plate V., Fig. 2), the follicles are almost always seen cut across; this, in my opinion, signifies that they have a sinuous form, and that they also, like the utricular glands, have an oblique direction. In a few cases, only, do we succeed, by vertical incisions, in cutting these follicles vertically also; they appear then under the form of extremely thin and irregular clefts or fissures. They differ essentially from the utricular glands in that they are very much shorter and more attenuated, and because their internal epithelium is paved instead of being cylindrical, as in the utricular glands (Plate V., Fig. 1, *a*). They both have one feature in common, that is, they increase in volume during pregnancy. In length, as well as in volume, the follicles vary sensibly among themselves. Those which are found agglomerated in the rudimentary cotyledons, and which are only the pedicle of the maternal cotyledon in the gravid uterus, as well as the utricular glands which traverse it, probably pour out the fluid secreted at the base and in the interior of the maternal cotyledon, or glandular organ of new formation. In both it is very difficult clearly to distinguish the orifice. Near the base of the glandular organ, segments of these follicles are often visible (Plate VI., Fig. 1, *m*), and what is most remarkable is not so much the increase in volume as the change in the secretion. In fact, their internal surface is no longer covered with pavement epithelium, and we discover upon it very diaphanous oval cells. It is difficult to say what their office may be; yet their increase in volume during pregnancy, their agglomeration in the cotyledons, and their change of secretion lead us to suppose that they are charged with an important function analogous to that of the utricular glands. At all events, what I have said upon the simple glandular follicles of the uterus of the cow, and especially those of the places where the placenta is formed, is sufficient, I think, to exclude the doctrines of those who taught and believed in the existence of short or simple glands in the uterus of animals, attributing to them the very simple function of secreting the uterine mucus.

In the centre of the uterine depressions which constitute the rudimentary cotyledons in the sheep we may observe, as I have said, numerous simple glandular follicles. The distribution of these glands differs in the sheep, however, from that which we find in the cow. In the rudimentary cotyledons of the non-gravid uterus of this latter animal the openings of the utricular glands are intermingled with those of the follicles (Plate V., Fig. 2); while in the rudimentary cotyledons of the sheep the utricular glands open around the elevated border which circumscribes the cotyledons, and in the central depression or concavity we see only, in very great numbers, the simple follicles.

I have several times stated that the utricular glands, as well as the follicles, increase in volume during pregnancy; but though this observation may be traced back, as far as the glands are concerned, to Malpighi himself, the measure of their increase I have not found indicated by any one.

The illustrious Professor Panizza[1] injected the utricular glands through their orifices, and he says that they are formed by a slender canaliculus, which is divided into two or three other smaller canaliculi, serpentine and goitre-like, terminating, after a course of three or four lines, in *culs-de-sac*. But he does not say whether he made the injections with the gravid or non-gravid uterus. It is very probable that it was during gestation, when the increase in volume of the

---

[1] Sopra l'Utero d'Alcuni Mammiferi, page 10. Milan, 1866.

glands renders the operation more easy; this, however, we cannot assert positively.

In the cow, the glands vary in length and in breadth in the gravid as well as in the non-gravid uterus, and the difficulty experienced in removing them entire and in sufficient numbers has prevented me from establishing an approximate measure of their length. I have therefore contented myself with noting the differences to be met with in measuring their transverse diameter, the sections in this direction being made, as I have said, quite easily both in the gravid and non-gravid uterus. I have always tried to measure them near their opening into the cavity of the uterus, because at that point we do not find the total or partial dilatations which are met with along the course of the uterine glands of certain animals, as in the cow, and which would be a source of error.

The following are the results which I have obtained: In the non-gravid uterus of the mare, the transverse diameter of the utricular glands, the thickness of the walls included, varies from 0.04 to 0.05 millimeters. In the uterus of the mare at term I have found the diameter to be from 0.05 to 0.06 millimeters.

In the gravid uterus of the cow, these glands reach a much larger development. For greater exactness, I took, in this animal, measures of the transverse sections upon the cotyledon of non-gravid cows, and upon the pedicle of the cotyledon of gravid cows. In the first case, the transverse diameter, the wall included, was from 0.09 to 0.10 millimeters; in the second case, in the third month of gestation, it increased from 0.14 to 0.16 millimeters.

I further measured the simple follicles from the same localities. In the non-gravid animals, the diameter of the follicles being, as we have seen, normally very variable, it changed from 0.02 to 0.04 mm.; in the gravid ones, from 0.04 to 0.08 mm.

We shall see, further on, how and why the internal cavity of the uterus in woman and in the mare ought, strictly speaking, to be considered as covered with a mucous membrane. At present we can admit the universal opinion that the uterine mucous membrane presents remarkable differences in the different species of animals. This being granted, we shall be permitted to quote the peculiarities we meet with in the uterine mucous membrane of rabbits, in which certain learned anatomists, like Bischoff, are not sure of the existence of the utricular glands, while others, like Leydig, have denied their existence in some of the rodents, as the rat, for example.

The mucous membrane of the non-gravid uterus of the rabbit appears to be formed of very delicate follicles, which might be called mucous, pressed close together, having a very narrow cavity or canal, varying in length from 0.05 to 0.08 mm. It was, perhaps, these follicles that Leydig, as I have before stated, compared with the intestinal glands of Lieberkuhn in the mole, without finding any essential difference between the latter and the very elevated folds that are visible in the uterine mucous membrane of the rat.

The opinion of Leydig receives strong confirmation when we examine comparatively the uterine mucous membrane of a gravid rabbit and the portions where no placenta has been formed. Here we meet large projecting linings or folds of the mucous membrane, where the glandular follicles that I have mentioned have really acquired a colossal development.

I shall speak of this more at length in treating of the formation of the placenta. For the present, I will simply observe that in examining the uterus of a pregnant rabbit, in which the upper portion of one horn had remained empty for about four centimeters of its length, I found the mucous membrane of this portion of the uterus covered with follicles of from 0.02 to 0.03 mm. in length, and 0.04 mm., or a trifle more, in width. They all had a much larger orifice, varying from 0.04 to 0.08 mm. In the largest, the opening, or central cavity, was 0.02 mm. in diameter; the thickness of the wall and of the internal epithelial layer measured 0.03 mm.

It was easy for me to ascertain these facts by lifting and detaching folds of the mucous membrane in the manner I have explained. By this means, we are assured of a very important fact, namely, that there exist no utricular glands in the uterus of the rabbit; while, from the vertical sections of the gravid and non-gravid uterus of this animal, there always necessarily remains something of uncertainty and doubt. We do, indeed, by the aid of such sections, easily see glandular tubes, but they are cut transversely and a little removed from each other (Plate VII., Fig. 4, *d, d*).

Without the method I have described, it was impossible to form an idea of the length and distribution of the so-called glandular tubes, which seem very few and scattered, when they are really only portions of the bases of the follicles, divided transversely, and included in the cutting of the preparation.

This explains the doubts of some, and the contradictions between those who maintained and those who denied the existence of the utricular glands in the uterus of rabbits. But what is of more importance is the further demonstration that the uterine utricular glands are really wanting in some of the mammalia. In that case, the mucous crypts or simple glands increase in volume during pregnancy in a truly extraordinary manner, and might lead us to suppose, with some appearance of reason, that they fulfill the function which belongs, under such circumstances, to the utricular glands and the simple follicles, when those organs are not lacking. From the very fact of the absence of the utricular glands, one can argue the great importance which they have during the period of gestation; but it is always very difficult to say precisely what their office is.

The numerous facts which I have just quoted give strong probability, I think, to the opinion expressed with fewer data by Eschricht, and adopted by several illustrious physiologists, namely, that these glands, to which I shall add the simple follicles, elaborate, during gestation, a fluid destined to furnish some elements for the nutrition of the fœtus, particularly when the organs which are charged with that office, or the villi of the chorion, are not fully developed. That both produce, in great measure, in the time of non-gestation, the fluid known under the name of uterine mucus is very probable; but what I emphatically assert is that there do not exist in the uterus of mammals two species of glands, distinct in form, volume, and office, as many authors have advanced.

In speaking of the human decidua, I shall show

that it may be considered as a product of exudation, due for the most part to the utricular glands, and that the numerous openings which perforate it obliquely precisely indicate the mouths of those glands, which remain open in the decidua for the continual passage of the materials elaborated by the glands. The same thing, though with entirely opposite results, is clearly seen in the fœtal envelopes of the cow, most anatomists denying the existence of the uterine or true decidua in that animal. It does exist, however, and it is easily seen in preparations hardened in alcohol or chromic acid. It is met with in the form of a yellowish pellicle, composed of epithelial cells irregularly stratified and covering the whole external surface of the chorion. The principal distinction of this epithelial layer, which represents the human decidua, is, that instead of adhering to the internal surface of the uterus, it adheres to the chorion, towards which is directed the product of the uterine glands. The result is that, instead of the cavities or sluices which we perceive in the human decidua, we find in the decidua of the cow, at the places corresponding to the uterine glands, small scales in the form of lids (of a yellowish, transparent substance, hard, and resisting the action of acids and alkalies), which infiltrate into the chorion itself, and there become assimilated.

Burkhardt[1] was the first to speak of this, and no

[1] " Quum igitur uterum a chorio removissem, propius lente vitrea armato oculo inspexi, et corpuscula minima lutea chorio inhærentia ex oculis vasorum spiralium recedere observavi. Simulæ ex vasorum orificiis hæc corpuscula remota fuerunt : guttulæ humoris lactei ex vasibus spiralibus extillaverunt." (Observationes Anatomicæ de Uteri Vaccini Fabrica, page 24. Basle, 1834.)

one has repeated his observation. Now, on examining these scales, I have been able to remark that they are formed from a thin and larger lamina which rests upon the chorion, and that on the part facing the uterus, opposite to the orifice of the glands, there is upon the lamella an ovoid protuberance of the form of a large nut, sometimes smooth and with yellowish central granulations, sometimes irregularly ridged and granulated. The size of these solid scales or lids is variable; some are rounded, and others oval or elliptical. The total diameter of those which are rounded exceeds somewhat 0.10 mm.; and the central oval elevation is 0.06 mm. in its greatest diameter. Among those of an elliptical form, the largest measured 0.30 mm. in their main length, and 0.16 mm. in breadth. The central elevation was 0.24 mm. in length, and 0.10 mm. in breadth.

What I have said thus far will suffice for the uterine glands and their functions. But it remains for me to speak much more at length of the glandular organ of new formation in the uterus of the mammifera and of the human species, which always forms, as I have already said, the maternal portion of the placenta.

CHAPTER III.

ON THE GLANDULAR ORGAN, OR MATERNAL PLACENTA, IN ANIMALS WITH A VILLOUS OR DIFFUSED PLACENTA.

THE ancient writers upon comparative anatomy maintained the absence of the placenta and of the cotyledons in the gravid uterus of certain animals, such as horses and swine. Modern writers have added that the villous or diffused placenta, as it is now called, is also met with in the camel, the dromedary, the llama, and, according to Müller, in the cetaceans.[1] A new comparative study of the gravid uterus of these animals may lead to the discovery of quite interesting anatomical differences.

It was long ago remarked that in the sow the papillæ (villi) of the chorion, instead of being short, very close together, and almost regularly distributed over the whole surface, are united in the form of pencils or tufts, the papillæ of the intervening spaces being much more thinly scattered. This condition constituted, it was thought, a transitional form between the diffused placenta and the agglomerated placenta of the ruminants.[2] For the time being, I shall examine only the villous or diffused placenta of the mare, not having been able to extend my ex-

---

[1] Müller: Manuel de Physiologie, t. ii., p. 731. Paris, 1851.
[2] Colin: Traité de Physiologie comparée des Animaux domestiques, t. ii., p. 560. Paris, 1856.

amination to that of other animals. Fortunately, it is claimed that the placenta of the mare presents a 'greater simplicity than that of the sow, and it should consequently be selected as a type for the observations I am about to offer.

On comparing the opinions that the science of the ancients has bequeathed to us upon this subject, we find that the teaching nearest the truth is due to our own Ruini. In the description of the placenta or secundine of the mare, he expresses himself as follows:—

"Upon these fœtal envelopes exists the red flesh, spongy and thin, which is met with in the matrix and the horns of the uterus of gravid mares. It is formed by the first blood of the uterine veins, which emerges almost thickened and curdled from their open extremities over the internal surface of the uterus. Its form is that of the uterus itself, because it is disseminated over the whole surface of it. It differs from the form of the placenta or secundine in woman, the latter being round like a cake."[1]

As a result of my observations, I am persuaded that these words of Ruini relate really to the glandular organ, which I shall soon describe, and which is formed over the whole interior surface of the uterus.

I would not, however, deny that by the words "carnaccia rossa e spugnosa" he may have intended to speak of the villi of the chorion, which were afterwards considered by authors as constituting, in themselves alone, the placenta of the mare; or, he may have referred to these villi and to the glandular

[1] Dell' Anatomia e delle Infirmità del Cavallo, Lib. IV., cap. xii. (Anatomia). Bologna, 1596.

organ or maternal portion of the placenta both together.

Eight years after Ruini, the celebrated Fabricius of Acquapendente, in his memorable work De Formato Fœtu,[1] considered the villi of the chorion as representing the placenta in the mare. This opinion was known and accepted by the illustrious Albert Haller, who, in the development of the idea, asserted that "in omnibus omnino quadrupedibus chorion repetitur, etiam in iis, quibus vera placenta vix tribui potest, ut in sue, ut omnino videatur naturam quadrupedum posse placenta carere, chorio carere non posse."[2]

Fabricius had also made another observation, which deserves special mention, both because he published it for the first time, and because it was afterwards either forgotten or repeated in a vague and uncertain manner. The observation is this: The small projections formed by the villi of the chorion enter into as many corresponding cavities of the uterus: "Minima et innumera tubercula per chorion dispersa et sese quasi mutuo tangentia, quæ in cavitates, sibi in utero respondentes, intrant, quæ sane carneæ substantiæ vicem subeunt."[3]

Now, if we seek for the developments of this discovery, from Fabricius to our own time, we shall find nothing satisfactory; for it has been either simply repeated or entirely forgotten.

"The chorion of the pachyderms," writes Baër,

[1] Padua, 1604.
[2] Elementa Physiologiæ Corporis Humani, t. viii., p. 185. Berne, 1766.
[3] Hieronymi Fabricii ab Acquapendente: Opera Omnia, Anatomia et Physiologia. De Formato Fœtu. Pars 11, page 89. Leyden, 1737.

"is covered with villi, which do not increase very much in size; they correspond to innumerable depressions upon the internal uterine surface, disposed like the cells of a honeycomb, each of which receives a villus of the chorion."[1]

Some years afterwards, Müller, instead of adding anything clearer and more precise, was even more vague and uncertain than Baer. He says[2] that in the pachyderms the fœtal placenta extends over the entire surface of the chorion, and that the maternal placenta is also distributed over the internal face of the uterus, which acquires a cellular texture and presents numerous depressions destined to receive the chorial villi.

My unsatisfactory retrospective study does not stop here. The ancients knew that the union of the chorion with the uterus, in the mare and the sow, takes place only during the last months of gestation. Yet not only was this observation forgotten (and it assumes to-day a very great importance in connection with the new formation of a glandular organ over the whole internal face of the uterus, which I am about to demonstrate), but things have been taught which are directly contrary to truth.

The first who drew the attention of anatomists and physiologists to this highly important fact was the Englishman Needham,[3] who, after having affirmed that in the pregnant sow the chorion adheres in no manner to the uterus, up to the middle period of

---

[1] Untersuchungen über die Gefässeverbindung zwischen Mutter und Frucht in den Saugethieren. Leipzig. 1828.

[2] Manuel de Physiologie, t. ii., p. 732. Paris, 1851.

[3] Disquisitio Anatomica. De Formato Fœtu, page 177, etc. London, 1767.

gestation, and that towards the end the projections or tubercles of the chorion adhere to it slightly, adds : —

"Equa quoque, ut sepius innui, prioribus mensibus fere eodem modo se habet et utero nusquam cohæret. Donec post aliquod tempus tubercula carnea exigua appareant orobi magnitudine. Hæc paulatim augentur, invicem continuantur et digitulos (non glanduloso corpore utero adnascenti[1]) sed ipsi uteri membranæ interiori, satis insignes inserit. Ut revera continuata quædam placenta per totum chorion extensa videatur, vel potius chorion ipsum ex membrana in placentam mutatum." And a little further on, at page 181: "Tandem in posterioribus mensibus eo ventum erit, ut chorion, notabili jam crassitie insigne, placentam utero continuatam repræsentet, surçulosque infinitos totidem venulis turgidos, uteri tuniceæ interiori immittat."

Snape[2] observed further that the chorion of the mare adheres to the inner membrane of the uterus only towards the end of the sixth month, and that, in the last months of pregnancy, the villi of the chorion so augment in volume that it seems to have lost the appearance of a membrane, and to have become a placenta.[3]

Wepfer[4] confirmed these observations, accrediting them to Graaf. But that celebrated anatomist really

[1] In speaking of the sow he had said : "Nullæ hic glandulæ, nulla placenta." (Loc. cit.)
[2] Anatomy of the Horse. London, 1606.
[3] This quotation was taken from the French translation of the work of Snape, by Garsault : L'Anatomie générale du Cheval, page 32. Paris, 1732.
[4] Ephem. Natur. Curios. Dec. 1, An. 1. 3 obs., p. 129.

did nothing but repeat observations already published by others, as we have seen.[1]

Haller afterwards coördinated,[2] so to speak, these scattered observations, but without adding anything of his own, and even leaving uncertainties, which must be kept in mind. Thus, after having said : —

"Qui equo et sui nullam placentam esse docent, ii prima tempora fœtus sola describunt, in quibus sola chorii membrana uterum sublinit. Nam etiam in his animalibus placenta sensim subnascitur, et in equo quidem chorion ex vasis nunc numerosissimis congesta in unam continuam placentam abit quæ cum utero confervet."

He adds a little further on : [3] —

"Quæ enim animalia eam conjunctionem [the union of the uterus with the placenta] habent leviorem iis etiam plus de chorii natura manet, et minus cum placenta humana convenit, ut equo et sui, et excusari possunt veteres, qui placentam pro crassiori chorii particula habuerunt."

I will not deny that the old writers were excusable, but it seems to me that this illustrious physiologist has simply repeated what they had said. All of them thought that the placenta was formed by the villi of the chorion. The same uncertainty, although more marked, can be traced back to Fabricius, who, with antique simplicity, confesses his own ignorance : —

---

[1] "In suibus vero per totam gestationem placenta nulla deprehenditur, at tantum chorii crassities quædam apparet: idem quoque in equibus primis gestationis mensibus observari dicitur, donec post mediam gestationem in chorio exigua tubercula carnea excrescant, quorum beneficio illud utero cohæreat." (R. Graaf: Opera Omnia, page 207. Leyden, 1678.)

[2] Idem, Physiol., t. viii., p. 233. Berne, 1766.

[3] Idem, page 226.

"In porcis autem et equis, quibus carnea moles nulla conspicitur, quid dicemus? Non certe quod aliquando Arîstoteles præcepit ut rem obscuremus, cum ignoramus: sed magis cum aliqua concinna sententia ignorationem tegamus: quæ est, ut in porcino et equino fœtu carneam substantiam ut in cæteris non observemus, quoniam Deus providentiam et potestatem suam multarum rerum mirabili varietate voluit ostendere."

A little further on he says also: —

"At cur equinus et porcinus fœtus ea destituantur, nihil habeo quod asseram. Nisi forte dicamus, non prorsus deficere, cum exterius per totum chorion exigua, imo minima, innumeraque tubercula, quasi se tangentia conspiciamus, cavitates illas in utero respondentes intrantia. Cur vero in equino et porcino fœtu ita habeant, explicet ille qui me felicior tantarum potuit rerum cognoscere causas."[1]

All these observations upon the gravid uterus of the mare, which to-day may be called ancient, have not, since their coördination by Haller, been developed as they deserved to be. Succeeding authors and the moderns content themselves with the statement that the uterine mucous membrane of the pregnant mare is tumefied; or, at most, like Baer and Müller, reproduce under a new but incomplete aspect what Fabricius and Needham had more clearly stated. None of them has ascertained or shown the real nature of the "carnaccia rossa" which was formed over the whole internal uterine surface, as was first pointed out by Ruini.

The moderns, in general, have admitted the idea

[1] Hieronymi Fabricii ab Acquapendente, op. cit., page 89.

that the villi of the chorion constitute of themselves the placenta of the pachyderms, disseminated or diffused over the whole surface. The opinion, incorrect indeed, of the celebrated Gurlt,[1] that in those animals the chorial villi penetrate directly into the utricular glands of the uterus, did not serve to change this notion.

If the opinion of Gurlt had been correct, it would at most have confirmed, as to those animals, the observations of Sharpey and Weber upon the penetration of the extremities of some of the villi into the first portion of certain uterine glands, where the placenta is formed.

I need only quote a few words of Colin to show how entirely forgotten in our own day have been the anatomical truths which I have just presented to you, as taught, though incompletely, by ancient writers:—

"The mode of union of the placental papillæ (villi) with the uterine mucous membrane is nearly the same in all animals, whatever may be the form of the placenta. On the uterine surface the mucous membrane presents, according to the careful observations of Berres and Weber, several kinds of follicles: some, broad and shallow, seem destined for the secretion of the mucus; others, very large and with numerous ramifications, destined to receive each a placental papilla and its various filaments. These follicles have, among the ruminants, enormous openings."[2]

[1] Handbuch der vergleichenden Anatomie der Haussäugethiere, page 431. Berlin, 1868.
[2] Traité de Physiologie comparée des Animaux domestiques, t. ii., p. 561. Paris, 1856.

Unfortunately, the conclusion of these historical researches is not encouraging. After more than two centuries, what little of truth was known has been forgotten; while new observations have been published, either untrue or imperfect, which have led us to a series of errors that I have summed up in the short quotation from Colin.

The comparative examination of the mucous membrane of the gravid and non-gravid uterus of the mare is of the greatest interest to anatomists and physiologists, because it reveals in all its simplicity and with clearness the constant double structure of the placenta. On comparing the mucous membrane of a non-gravid uterus with that of a gravid uterus almost at term of this species of animals, we remark, even by simple external and superficial observation, certain differences of form and color.

In the non-gravid uterus, the mucous membrane is smooth and velvety, of a rose color, inclining to yellowish; here and there it is doubled over into large and soft folds. In the gravid uterus, on the contrary, the folds are wanting, and over the whole surface of the mucous membrane are scattered vermiform projections, winding closely against each other, of a very intense, deep red color, tending to violet; the surface, instead of being smooth, has a swollen appearance, and, as we observe it, we cannot help thinking of the "carnaccia rossa e spugnosa" of Ruini.

On cutting vertically portions of the gravid uterus at an advanced period of gestation, and comparing them with similar sections in the non-gravid uterus, we perceive, even with the naked eye, that the mucous membrane, which is scarcely distinguishable in the latter, is considerably enlarged in the former, and that it constitutes a uniform layer, varying from a millimeter and a half to two millimeters in thickness, of a yellowish-red color, and on the interior of the cut surface having a swollen appearance. By carefully disengaging the chorion, we easily succeed in seeing that the numerous and closely-packed villi which cover it, enter into the thickened layer, into which the mucous membrane seems to have been transformed.

It is extraordinary that very respectable anatomists and physiologists should have contented themselves with these simple and superficial observations, and not have carried their researches further. I have until now added nothing of my own to what Fabricius and Needham had taught, and von Baer and Müller had repeated with somewhat less completeness. But microscopic examination quickly revealed to me the easy and positive explanation of these transformations, which I shall now briefly point out.

In Plate III., Figs. 2 and 1, I have represented vertical sections of the gravid and non-gravid uterus of the mare, very slightly magnified, in order to show, at first sight and a little more clearly, the differences which, as I have already said, are visible even to the naked eye. The dark line (Fig. 1, *a, a*) represents the epithelial layer which covers the whole internal surface of the non-gravid uterus in the mare. The letters *b, b* indicate the utricular glands; some are entire, while others are cut in different directions. The magnifying power employed is so slight that we cannot distinguish any element of the sub-mucous, cellulo-vascular layer, in which the glands are imbedded.

Fig. 2 represents a like vertical section of the gravid uterus in the same animal; and what first strikes us is the increased size of the line marked *a a*, in the preceding figure. It is no longer a dark line, in which the cellular elements that compose it are not apparent, but a layer uniformly undulating, and formed by masses resulting from the union of small sacs or follicles pressed closely together (*a, a*). The masses are separated from each other by the secretory canals of the utricular glands (*b, b*), which, by reason of the thickness of the sub-mucous layer and their own increase in volume, are more easily cut transversely in the gravid uterus.

From these few features that I have pointed out, there arises a two-fold and very important question : What is this uterine mucous membrane in the mare, and what portions ought to be comprised under that denomination ? What parts of the mucous membrane undergo transformation, and thus give rise to the differences we have remarked ?

In order to determine of what this mucous membrane consists, I must first of all recall the fact, that for a long time anatomists were uncertain whether the internal surface of the uterus in woman was or was not covered with a mucous membrane. Bischoff, in his treatise upon the development of the ovum, touched upon this question in speaking of the membrana decidua, and he says : "If the admission of a mucous membrane in the uterus of woman implies that it may, by the scalpel, by maceration, or by any other operation, be distinguished and separated into an internal layer, special and membraniform, as in the majority of the mammals, it must be allowed that the uterus of woman has no mucous membrane; for even with the thinnest vertical sections, and with the aid of the microscope, there is nowhere visible even the trace of such a layer distinct from the uterine parenchyma."

Then he adds: "If we consider the nature of the internal surface of this organ, we see that it has strong resemblances to a mucous surface."

It seems to me, to tell the truth, that Bischoff, in this way, has evaded rather than resolved the question. And what has been said concerning the uterine mucous membrane of woman may be also asserted in reference to the gravid or non-gravid uterus of the mare. With regard to this species of animals, we should fall into even greater contradictions; for in deciding that the uterine mucous membrane in the gravid mare is formed of the layer of glandular follicles, of which I have before spoken, it would also be necessary to affirm that in those animals the uterus is covered on its internal surface with a mucous membrane during gestation only, and that it is wanting in the non-gravid uterus, unless we chose to give the name of mucous membrane to a simple epithelial layer, and then the uterine mucous membrane would exist in the mare when not pregnant, and would be lacking during pregnancy. Now, the peculiarities that I have mentioned as existing in the uterine mucous membrane, when treating of the glands of the uterus, will enable me, I think, by again referring to them, to answer this question, which has some interest for anatomy as well as for physiology, and which has a special interest for me, forced as I shall be, constantly, to speak to you of the mucous membrane of the uterus.

The solution of this problem, at first sight very complex and difficult, if we limit our researches to the human uterus, becomes, on the contrary, very easy, in my opinion, if we call to our aid comparative anatomy. Indeed, what composes the uterine mucous membrane in the animals in which it is admitted by all without dispute? It is claimed that the mucous membrane exists when, without the aid of any method of dissection, a membraniform layer can be easily lifted from the internal surface of the uterus; it is claimed with yet more assurance when this layer rises under the form of folds or plaits, more or less apparent. But appearances should not be confounded with reality.

In examining, with the aid of the microscope, the internal surface of the uterus of the mare or of woman, where a true mucous membrane, in the anatomical sense of the word, does not exist, and a festooned fold of the uterus of a rat, in which the existence of the mucous membrane is not disputed, the observer notes no real difference between their histological elements.

It has already been said that in the mare it is an epithelial layer which rests upon a soft connective tissue of the internal uterine surface, while in woman the same thing is observed, with this sole difference, that the connective tissue is more dense, more compact, and more adherent to the epithelial layer. But in the rat the same thing holds true also, with this one peculiarity, that the connective tissue rises from the surface of the uterus, and produces fringes and festoons of the mucous folds. This is so true that on removing the connective tissue of one of these folds, so as to loosen the folds and festoons, we find the identical constitution of the parts which form the mucous membrane of the uterus of those animals in which, as well as in woman, the existence of this membrane is denied. I have already mentioned the fact that in the epithelial layer which covers the internal surface of the non-gravid uterus in the rabbit we see small, shallow depressions, or very simple glandular follicles.

Such are the principal anatomical differences of structure in the uterine mucous membrane that I have been able to detect in the animals I have examined, or that I have found to be indicated by the most celebrated anatomists. But, instead of denying, after the study of these facts, its existence in woman and in certain animals, and admitting its presence in others, it seems to me more logical and more in conformity with the truth, to affirm that the simplest form of the mucous membrane is represented in woman and in the mare by an epithelial layer; that its structure does not change essentially when the sub-epithelial connective layer is more or less soft, or when it rises above the level of the internal surface of the uterus in the form of folds and festoons; and that, further, the greatest apparent complication to be observed in this membrane is found in the epithelial depressions which we remark in certain cases, but these, even, lose their seeming importance when, with good reason, as I have tried to demonstrate with Leydig, we consider the large folds of the mucous membrane as colossal glandular follicles. Hence follows, as a consequence, the confirmation of the idea that a pellicle (*velamento*) represents the simplest and most fundamental form of the mucous membrane which lines the inner surface of the uterus of all mammals, the human species included, and that this form is modified, without radical change, in the distinctive appearances observed in the different species of animals.[1]

Thus, it seems to me, is more clearly defined the somewhat vague opinion of the majority of anatomists, who, while admitting a mucous membrane in the uterine cavity, have claimed that it is inseparable from the subjacent cellulo-vascular tissue, and not perceptible even with the most powerful means of microscopic observation.

This, strictly speaking, amounts to the assertion, "It exists, though we are not able to see it."

The uterine mucous membrane, compared by Bischoff with other mucous membranes of the animal structure, from a physiological point of view and in relation to other functions, presents differences much more important and much more essential, physiologically speaking, which distinguish it from all other mucous membranes. The most fundamental of all, is the marvelous transformation which it undergoes, either at a fixed point, or over its whole extent, during pregnancy, in order to give rise, always and under all circumstances, to the neoformation of a transitory glandular organ, constituting the maternal part of the placenta.

I shall now limit myself to the explanation of what

[1] See Appendix.

takes place over the whole internal surface of the uterus of the mare during gestation, and I will begin with the exposition of this very important phenomenon when fully developed, that is to say, when pregnancy has arrived at term.

The part which I have heretofore said had a tomentose appearance, the "carnaccia rossa" of Ruini, or the tumefied mucous membrane of the moderns, with its numerous depressions, is no other than a glandular organ of new formation, constituting in the mare the maternal portion of the placenta. In Plate IV., Fig. 1, I have represented an entire follicle, from a vertical section of this glandular organ, which is in connection with the villi, the villi being also connected with the chorion. In Plate V., Fig. 1, I have reproduced a transverse-oblique section of the same glandular organ, to exhibit more clearly its intimate structure.

The new glandular organ, which arises from the sub-epithelial connective tissue of the whole surface of the uterus, is formed by the aggregation of an infinite number of simple glandular follicles, of which a few only have a double and even a triple *cul-de-sac* (Plate V., Fig. 1, *a, a*).

The type of these follicles seems to be single as to form, and the bifurcation of the base of some among them appears to arise from the blending of two of these follicles which are pressed close together. Their average length is, as I have already stated, from a millimeter and a half to two millimeters. The diameter varies according as the measurement is made at the base or at the orifice, for they are pyriform. At the apex, or a little below the orifice, which resembles a small tunnel (Plate IV., Fig. 1, *i*), they measure three one hundredths, towards the middle portion four or five one hundredths, and at the base from eight to twelve one hundredths of a millimeter.

The sub-mucous connective tissue proliferates and is found interposed between the follicles, so as to furnish a wall to the external membrane of each one of them. In proximity with the utricular glands, the connective tissue develops more abundantly in the form of interposed pyramids (Plate V., Fig. 1, *c, c*), and it accompanies the glands as far as to their opening into the internal surface of the uterus, and precisely to the plane where we find the orifices of the follicles of new formation. From these pyramids diverge as many septa, which communicate with the connective tissue surrounding each follicle. In the centre (Plate V., Fig. 1) we see one of these pyramids cut transversely, in the midst of which appears a venous vessel (*d*), and two utricular glands, also cut transversely (*f, f*).

The whole internal surface of the glandular follicles of new formation is covered with pavement epithelium, which is more easily distinguished in the transverse sections (Plate V., Fig. 1, *b, b, b*) than in the longitudinal ones (*a, a*). From the uterus extends a rich vascular net-work, which has a very peculiar arrangement. From several trunks (Plate IV., Fig. 1) ramify smaller trunks or loops, assuming the form of a brush, which traverse the spaces between follicle and follicle from their base to the extremity (*l, l*). From the stems of these loops diverge other lateral vessels (*m, m*), which anastomose with each other, and go to make up the dense vascular net-work which surrounds each follicle. The net-work is so thick that it is not rare to find entire portions of it in the transverse incisions of the glandular organ.

In Plate V., Fig. 1, *e, e,* the vessels we have just spoken of represent the utero-placental vessels of animals which have a single placenta, and even those of the human species.

Each villus of the chorion penetrates and fills a follicle (Plate IV., Fig. 1, *c*). It is formed of a vascular loop, ordinarily simple, sustained by the soft connective tissue which is furnished it by that of the chorion (*f*). All the villi are surrounded by an epithelium (*d*) which is continuous with that which covers the whole exterior of the chorion (*b*), the outer surface of which may be regarded as representing the uterine decidua in the mare.

In the same figure (*a, a*) we see the chorion formed of corpuscles of connective tissue, in the midst of which traverse the vessels (*c*) destined to form the umbilical cord.

The utricular glands of the uterus pour out their fluid between the glandular organ and the external surface of the chorion, where there is collected in abundance a whitish, albuminous liquid, resulting from a mixture of the above fluid with that which is secreted by the follicles of the glandular organ.

At what period of gestation begins the formation of this new glandular organ, which represents the uterine or maternal portion of the placenta in the simplest manner and under the most simple forms by its general diffusion over the whole internal surface of the uterus? I cannot answer with precision.

However, it is certain that the observation made by the ancient anatomists, that the villi of the chorion adhere to the uterus of this animal only during the last months of pregnancy, leads us to suppose, with reason, that this assertion should to-day be understood to signify that the glandular organ is formed late in the uterus.

I have only been able to observe a portion of the uterus of a mare during the early periods of gestation. It had been preserved in alcohol for several years, and I discovered in it no trace of a glandular organ. Consequently, it is impossible for me to affirm anything probable with regard to the time and mode of formation of this new glandular organ.[1] As a compensation, I was fortunate enough to be able to examine the uterus of a mare killed fifteen or twenty hours after the delivery of a fœtus at term. The glandular organ had remained entire in the uterus, It is certain, then, that the delivery of the animal is not accompanied by any traumatic lesion. The villi of the chorion come out from the follicles like the fingers of a glove. Yet the glandular organ had, after so short a time, undergone remarkable modifications. The livid red color had become yellowish; the volume had diminished by half; the follicles measured no more than a millimeter or a millimeter and a half in length; their internal diameter, examined by the microscope, was only a hundredth or one and a half hundredth of a millimeter at the extremity, and from four to five hundredths at the base. The vascular net-work between the follicles was no longer perceptible.

[1] See Appendix.

It also remains to be known how much time is necessary for the destruction of the glandular organ or maternal placenta, in what manner it is accomplished, and how the internal surface of the uterus is again covered with the epithelial layers which envelop it during non-gestation. I would gladly resolve these questions, and many others that relate to the period of formation and the mode of disappearance of this glandular organ or maternal portion of the placenta that I have described, in this too expensive animal.

Now that I have made known the most important part of the phenomenon, it gives me pleasure to believe that others, more fortunate, will meet with favorable opportunities for observing its special features.

CHAPTER IV.

ON THE GLANDULAR ORGAN, OR MATERNAL PLACENTA, IN ANIMALS WITH A MULTIPLE PLACENTA, AS IN THE RUMINANTS; OR, ON THE UTERINE COTYLEDONS IN THOSE ANIMALS.

In the analysis given of the anatomical knowledge which the ancients have transmitted to us upon animals with a villous or diffused placenta, we have seen that they had indicated or had glimpses of the truth, but that, in the course of centuries, it had been forgotten, and erroneous doctrines followed. With regard to the cotyledons, or multiple placenta, we shall see that errors were at first committed, which observers soon corrected, without, however, discovering the entire truth.

It is claimed that Diocles described the uterine cotyledons, even in woman; and that Hippocrates spoke of them in his aphorisms under the name of "acetabula uteri,"[1] an error for which he was reproved by Galen.

Aristotle corrected the error of Diocles, affirming that "dentata animalia cotyledones habent," and that "utrinque dentata non habent cotyledones."[2]

---

[1] Quæcumque mediocriter corpora habentes abortiunt, secundo aut tertio mense sine occasione manifesta: his "acetabula uteri" plena mucoris sunt, et non possunt ex pondere fœtum continere, sed disrumpuntur. (Aphor., Sect. v. xlv.) Several commentators have affirmed that Hippocrates, after the example of Proxagoras, had given the name cotyledons to the vascular orifices opening into the cavity of the uterus.

[2] Aristotle: Historia Animalium, cap. v.

Our countryman Aldrovandi was the first to give a diagram of the cotyledons of the cow,[1] but it is not a very good one. On the contrary, the illustrations produced by Hobokenius,[2] especially those of Plates XIV. and XV., are very beautiful.

Fabricius of Acquapendente [3] and Marcus Aurelius Severin [4] were among the first to remark the differences of volume which the uterine cotyledons undergo in the gravid and non-gravid uterus, particularly in the sheep and the cow.

"In non gravidis," wrote Severin, "quidem similes grano tritici, in gravidis vero corporis raritale foraminulenta similes hæ cotyledones sunt spongiæ candidæ." But Aristotle had long before pointed out their increase during pregnancy, and after delivery "minora reddantur, demumque obliterantur." [5]

Needham had added that the caruncles of the chorion at the beginning of gestation are detached with difficulty from the uterine cotyledons, but that with the development of the fœtus they separate easily, as if mature, " et sponte cum fœtu abeunt." [6] Then he continues : " Glandulæ vero ipsæ in utero relictæ paulatim decrescunt."

Hobokenius also called the cotyledons of the uterus glands; but he described better than did the others [7] (and for the first time, I believe, under the denomination of ligament of the uterine glands) the

[1] Aldrovandi, Ulis.: Quadrup. Bisulc. Historia. Bologna, 1621.
[2] Hobokeni, Nicolai: Secundinæ Vitulinæ Anatomia. Utrecht, 1672.
[3] De Formato Fœtu Pat. 1604.
[4] Zoönomia Democritea. Nuremberg, 1645.
[5] De Generatione Animalium, cap. v.
[6] Disquisitio Anatomica de Formato Fœtu, page 184. London, 1667.
[7] Secundinæ Vitulinæ Anatomia, page 143. Utrecht, 1672.

pedicle of the cotyledons in the gravid uterus. He pointed out and sketched, with great exactness, the uterine vessels which go from the so-called ligament to the glands or cotyledons of the uterus.

To this knowledge of the ancients Malpighi[1] added nothing; the moderns very little. It is generally taught that the uterine cotyledons in the ruminants are appendages of the mucous membrane; that they are perceptible in a rudimentary state, even in the uterus of the fœtus; that they acquire some development after birth, become hypertrophied during gestation, and are persistent throughout the life of the animal.

In describing the utricular glands of the uterus, I observed that Professor Spiegelberg had stated, very incorrectly, as we shall see, that the uterine cotyledons were nothing but expansions or dilatations of those glands.

An illustrious Italian anatomist, Professor Panizza, in his last published work[2] remarks, while treating of the cotyledons, that the mammillary elevations visible in the uterus of the heifer are only the rudiments of future internal cotyledons; and he speaks of the differences which they present in their development and size, according to age, the condition of gestation or

---

[1] Observantur quoque quamplurimi in tota uteri et cornuum interiori superficie tumores inæqualis magnitudinis parum assurgentes, qui graviditatis tempore insigniter surgent et uteri appendices videntur seu vaginularum congeries unde cotyledonum nomine insigniuntur. Admittunt autem erumpente a chorio subintrantes radices ita, ut ex his duabus insitis partibus completa habeatur glandula, qua separatum ab utero alimentum fœtui subministratur. (Opera Omnia. Epist. ad Sponium, page 29. London, 1667.)

[2] Sopra l'Utero gravido di Alcuni mammiferi, pages 11, 13. Milan, 1866.

non-gestation, and the place which they occupy. He accepts the comparison of the ancients, likening the uterine cotyledons to the esculent egg-plant, and describes the slender stem or peduncle which Hobokenius had pointed out under the name of ligament.

Panizza states that it is flattened, formed by the uterine mucous membrane, supplied with vessels of all kinds and with nerves belonging to the cotyledon, and that the alvooli of this lattter organ, more or less large and deep, are subdivided into secondary alveoli. In advance of all others, this anatomist has touched upon the important question of the development of the cotyledons during pregnancy. And if he leaves much, perhaps, that we may desire to see explained in connection with this subject (especially after I shall have demonstrated that the portion of the maternal cotyledon is not so much the product of a hypertrophy of the preëxisting cotyledons as a veritable neoformation of a glandular organ at the place corresponding to the mammillary tumefactions of the gravid uterus), I am yet happy to repeat his very words, because they mark the date of those new and interesting researches which ought to be still further extended in order to ascertain how, after delivery, this glandular organ of new formation is destroyed or lost.

Panizza writes thus:[1] —

" By following in succession the early periods of gestation in the cow, we discover how the maternal or fœtal cotyledons are developed. In the uterus of the cow, examined from the tenth to the thirtieth day after fecundation, we perceive the external membrane of the envelope of the fœtus, the chorion, in

[1] Op. cit., p. 13.

simple contact with the internal surface of the uterus, and at those places only which correspond to the future maternal cotyledons.

"The chorion becomes more opaque and full of small elevations, or soft, white points, more or less raised, according to the age of the embryo. Observed with the magnifying glass, these points appear more or less elongated and transparent. They are the rudiments of the cotyledons of the fœtus, simply supported upon the uterine parts which correspond to the rudiments of the cotyledons of the mother. As soon as it is recognized that the maternal cotyledons are only very soft expansions of the mucous membrane and of its corresponding small vessels; as soon as it is understood that the fœtal cotyledons are vascular projections of the vitelline membrane or chorion of the fœtus, it is then clearly seen how, subsequently, with the development of these two parts, the maternal cotyledon must present an entirely alveolar surface."

The old writers have bequeathed to us but few observations upon the differences of the cotyledons in the various species of ruminants. These are very incomplete, and we have simply accepted them. As far as their form is concerned, Fabricius had said what the moderns repeat with Needham:[1] "Ovis et capra, per omnia vaccæ similis est; præterquam quod glandes quæ illic convexæ sunt, hic concavæ apparent et cotylæ sive acetabuli nomen sensu maxime proprio ferunt."

We still agree with Harvey that the cotyledons of the hind resemble in form those of the cow, but they are much smaller and less numerous, since only five

[1] Needham, op. cit., pp. 185, 188.

have been counted in the hind, and more than eighty in the cow.

The differences are not limited to the form of the uterine portion, concave in one and convex in the other. But I shall occupy myself with these diversities at another time.

It has always been affirmed that the function of the cotyledons in the ruminants is to elaborate the nourishment of the fœtus, but the ancients, as well as the moderns, have never been able to agree as to the manner in which this important process is carried on.

Aristotle maintained that in the cotyledons [1] "veluti mamma reponitur a natura fœtui alimentum sanguineum."

Fabricius [2] admitted the direct communication of the maternal vessels with those of the fœtus in the cotyledons. He saw an argument in confirmation of this opinion in certain cells containing a black pigment, which are often observed in some of the cotyledons of the sheep: "Plurimis atrisque punctis, quæ ab ruptura orificia venarum sunt."

William Harvey [3] constituted himself the champion of the opposite doctrine: "In cotyledones alimentum fœtui reconditur non quidem sanguineum, ut Fabricius voluit, sed mucosum, ovique albumen crassius plane referens. Unde etiam manifestum est bisulcorum fœtus, ut alios omnes sanguine materno non ali."

Needham [4] was still more explicit than Harvey:

---

[1] Op. cit., loc. cit.
[2] Op. cit., p. 39.
[3] Exercitationes de Generatione Animalium, page 579. Padua, 1656.
[4] Op. cit., p. 25.

"Per molem carneam filtratur succus nutritius in placentiferis omnibus et in glanduliferis, sive ruminantibus. In ruminantibus hoc peculiare obtingit, quod succus, priusquam carunculas carneas chorio accrescentes ingreditur, in glandulosa corpora extuberat, quæ loculamentis quibusdam, quasi favoram alveolis ubique terebrata, surculos et digitulos a placentibus chorii exporrectos recipiunt, iisdemque se mulgeri sinunt."

Finally, not to quote too many names, I shall simply notice that Haller[1] formulated this categorical sentence : —

"In ruminantibus manifestum fit, matrem inter et fœtum, non sanguinis, sed lactis esse commercium."

The ancients also gave to the cotyledons the name of uterine mammæ; and several among them, before Haller's time, employed the term uterine milk to designate the fluid which is found in the cotyledons.

In speaking of the nature of this fluid, which he calls mucous albumen, Harvey says it was also known to Galen.[2] Vesalius declared it to be a mucous substance. Malpighi[3] demonstrated that in the process of cooking it all the characteristics of albumen, subjected to heat, were plainly detected. With Needham, the majority were content to call it uterine milk, and to qualify it as a fluid resembling real milk. Vieussens pronounced it true milk.[4]

Among the moderns, Duverney[5] and Eschricht re-

---

[1] Elementa Physiologiæ, t. viii., p. 296. Berne, 1766.
[2] Op. cit., p. 574.
[3] Opera Posthuma, page 162.
[4] Nov. Vas. Syst., p. 41.
[5] Œuvres anatomiques, t. i., p. 538.

garded the uterine portion of the cotyledons as a true gland, asserting that its fluid, absorbed by the villi of the chorion, serves as nutrition for the fœtus. Just here, we must not forget a comparison of Harvey's with regard to the nutrition of the fœtus in the ruminants during the different periods of intra-uterine and extra-uterine life: —

"Idque manifestum est, quod de cotyledonibus in cervarum aliorumque bisulcorum carunculis supra diximus: nempe carneam molem in iis animalibus spongiosam esse, et, favi instar, infinitis pene acetabulis constare, eandemque mucoso albumine repleri, atque inde vasorum umbilicalium fines nutrimentum haurire, quod in fœtum transferant: quemadmodum in jam natis anima libus venarum mesentericarum ramuli, per intestinorum tunicas diffusi, ex illis chylum absorbent."[1] .

Returning to the quality or chemical composition of the fluid found in the cotyledons, I will state that Prévost and Morin were the first to make an analysis of it, and they found it to be composed of albumen, fibrine, caseine, a gelatinous substance, a red coloring matter, osmazone, fat, and salts.

Schlossberger, of Tübingen, examined in 1855 the uterine milk of the ruminants, and he found that this fluid has the consistency of cream, and appears under the microscope to be composed of free nuclei, globules of fat and epithelial cells; it possesses a slightly acid reaction, and contains albumen and salts, but no sugar.

Dr. A. Gamgee[2] has since shown that the reaction

[1] Op. cit., p. 574.
[2] Edinburgh Veterinary Review, No. 46. Edinburgh, 1864.

of the liquid is alkaline, and he obtained the acid reaction only at the beginning of the process of decomposition. He has established the presence of water, albumen, alkaline albuminates, fat, and inorganic salts.[1]

Spiegelberg, of Freiburg, like Gamgee, found neither sugar nor caseine in the fluid, and he does not think proper to give to it the name of uterine milk.[2]

As a fact which may be of importance in explaining the diversity of these results, we must remember that Bernard,[3] in 1855, obtained sugar in the muscles and lungs of the sheep, the dog, the rabbit, and even of the human fœtus, during the first periods of intra-uterine life. He also demonstrated its presence in the liquid of the allantois, of the amnion, and of the urinary bladder. He maintains that the sugar disappears from these liquids, as from the tissues of the fœtus, just in proportion as the glycogenic function of the liver becomes established.

Three years after, he affirmed that he had sought

---

[1] The following are the results of the chemical analyses of Prévost and Morin and Dr. Arthur Gamgee.

Prévost and Morin out of 100 parts of fluid from the cotyledons of the cow.		Gamgee out of 1000 parts of fluid. Alkaline reaction. Specific gravity, 1.033. Fahr. 60°.	
Water	86.837	Water	879.10
Solid parts	13.163	Solid Parts	120.90
Albumen and fibrous sub.	11.028	Albumen	104.00
Gelatinous matter	0.546	Alkaline albuminates	1.60
Fat	0.750	Fat	12.33
Osmazone	0.714	Inorganic salts	3.74
Traces of salts	0.714		

[2] Zeitschrift für rationelle Medizin von Henle und Pfeufer, b. 21. 1864.

[3] Leçons de Physiologie expérimentale appliquée à la Médecine. Paris.

in vain for several years for the glycogenic matter in the cotyledons of the sheep and of the cow, during the different periods of their life. He also thought he could demonstrate that if in animals with a single placenta we meet, mingled together, the vascular and glandular portions which, in his opinion, secrete the sugar, the same parts are developed separately upon distinct membranes in the ruminants; that is to say, the vascular portion upon the chorion, and the glandular portion upon the internal surface of the amnion.

The glandular organ, or the glandular or glycogenic cells, would be formed, according to him, by the whitish plates that are met with, during the first months of pregnancy, upon the internal surface of the amnion, and whose physiological significance was entirely unknown, even by those who had been acquainted with them before his time.[1]

This subject awaits from chemistry a long series of experiments which shall interpret the phenomena in a positive manner, and clear up the difficult question of the nutrition of the foetus. I will simply state here, however, that in treating, either with nitric acid or by ebullition, the milky fluid of the cotyledons, which lubricated the internal uterine surface of a mare at term, I recognized the presence of albumen; with the tincture of iodine and the addition of a drop of sulphuric acid, I obtained the characteristic reaction of starch; by using iodine alone, the disturbance peculiar to dextrine was produced; finally, by a solution of nitrate of silver there were visible traces of the existence of chloride of sodium. In briefly

[1] Mémoire sur une nouvelle Fonction du Placenta. (Annal. des Sci. Nat., 4 Série, Zoölogie, t. x., p. 112. Paris, 1858.)

reviewing the different opinions held concerning the structure of the cotyledons and the composition of the fluid which they secrete, I ought also to mention that, in recent times, Colin[1] has affirmed that the so-called fluid which they contain is only the product of an illusion, that is to say, the effect of post-mortem decomposition. Notwithstanding the assertions of Colin, these observations are so easily made and so palpable that they have been universally supported, and contradicted by no one else.

The opinions which divided the ancients upon the functions of the cotyledons are then reduced to two. They are the same that still prevail to-day, modified and adapted, so to speak, to the language imposed upon us by the progress made in anatomy and physiology. The two chiefs of the school of antiquity are Fabricius, who admitted the direct communication of the maternal vessels with those of the fœtus in the cotyledons, and Harvey, who maintained that the fluid secreted by the cotyledons was absorbed by the villi of the fœtal placenta.

Among the partisans of Fabricius we may at present reckon all those (and they form the greatest number) who believe that the fœtus is nourished in the uterus by means of an exchange of materials in the cotyledons between the maternal and the fœtal vessels. In the number of the partisans of Harvey, we may rank all those who admit, for the ruminants at least, the absorption of the fluids separated by the cotyledons of the mother. To the latter opinion adhere all those who think that the fluid secreted by the uterine glands serves as nutrition for the fœtus,

[1] Op. cit., t. ii., p. 600.

and that the cotyledons are nothing but the dilatations or expansions of the mucous membrane, or of a portion of the glands. This opinion has been embraced by Spiegelberg, and was indicated by myself in speaking of the utricular glands in the ruminants.

One teaching of the older anatomists, with regard to the maternal cotyledons of the ruminants, has reached us entirely unchanged, notwithstanding all the modifications introduced into the description of the fact. It is that in the uterus of the ruminants, even at the fœtal period, we find rudiments of cotyledons, which in extra-uterine life increase in size, are greatly developed during pregnancy, and diminish after delivery, though always remaining in the uterus.

Nor has this teaching been essentially modified by the views of those modern writers who, instead of attributing the development of the cotyledons, during gestation, to an enlargement of the mucous membrane, ascribed it to a dilatation of a portion of the utricular glands. No one suspected that this development depended upon the formation of a new glandular organ, which differs from that I have described as being observed in the uterus of the mare, in one feature only, namely, that in the latter animal the neoplasm takes place over the entire uterine surface under the simplest forms, and not over certain circumscribed areas, with a more complicated structure, as in the ruminants.

Before making this important observation upon the uterus of the mare, which opened the path to the investigations of which I treat, I lost several months in the comparison of the structure of the maternal cotyledons of the gravid and non-gravid uterus of cows. My patient fellow-laborer in these fruitless researches was the excellent young doctor Severi. But we could never detect a single indication, however vague, of the probability, to say nothing of the certainty, of the doctrine supported by Spiegelberg.[1]

In speaking of the uterine glands, I stated that, by the aid of transverse sections of the cotyledons of the non-gravid uterus, I had assured myself that, besides the utricular glands, very numerous agglomerations of slender follicles were to be seen in them (Plate V., Fig. 2). I was very far from suspecting the neoformation of a special glandular organ in the uterus of all mammals, whatever might be the form of the placenta; and the observation did not allow me to admit either the dilatation or the expansion of the utricular glands in the formation of the cotyledons of the gravid uterus. I therefore sought for the solution of the problem in these slender follicles, which I supposed must become hypertrophied during gestation, and remain atrophied during the period of vacuity.

When I speak of the formation of the placenta in those animals in which that organ is single, and when I describe what I have seen take place in the small follicles of the uterine mucous membrane of rabbits, it will appear that it would have been an allowable supposition that the same phenomenon might be repeated, with certain modifications, upon the mucous membrane of the cotyledons, in order to form their glandular portion. I prefer, however, to remain in doubt until facts shall have settled the matter conclusively; all the more, that certain observations

[1] See the first part of this Memoir upon the Uterine Glands.

seem to contradict this method of formation for the glandular or maternal part of the placenta in the ruminants.

In making vertical sections of the cotyledons and the walls of a gravid uterus, we observe three distinct parts: first, the walls of the uterus; second, a fold of the mucous membrane, narrowed at its base, broadening towards the summit, whence arise pyramidal dissepiments of connective substance, which go to make up the third part, or superior portion of the cotyledons. The fold of mucous membrane constitutes that which is at present called the pedicle of the cotyledon, or the ligament of Hobokenius, through which the vessels are carried to the third and last part, that is to say, to the glandular portion of the cotyledon of ancient and modern anatomists.

The pedicle represents the old or permanent part of the uterine cotyledon of the cow. I do not speak of the differences which are observed, in regard to this, in the sheep. The third or superior portion, more remarkable still, represents the glandular organ of new formation, which remains in the uterus after delivery, and which then disappears, as we have seen occur with the uterine glandular organ of the mare. It remains to be demonstrated whether, in both cases, this complete disappearance is the effect of a simple progressive atrophy; or whether, in the cow, the glandular portion of the cotyledons is destroyed by fatty degeneration, as the uterine mucous membrane is, at the place where the single placenta of the carnivora has been formed. Yet, from the facts described above, it follows evidently and clearly that, as a result of pregnancy, the convex surface of the old cotyledon has lost the form and anatomical structure which it possessed. It is no longer smooth and slightly convex; for from its whole surface have arisen long and slender pyramids, which, in the sections, appear almost digitate, and consist, in a great measure, of connective tissue of new formation (Plate VI., d, d), and of vessels which are a prolongation of the uterine vessels (Plate VI., f, f). Then, as a consequence of pregnancy, there takes place in the fixed and permanent portion of the cotyledons a neoformation of connective tissue, which, we shall see, is the stroma of the new glandular organ. In short, there is produced, in a more complete and complex manner, what we have seen take place normally in the uterine mucous membrane of certain mammals, in which, the uterus being non-gravid, this membrane rises in large, festoon-like folds.

The examination of the pedicle by transverse or longitudinal sections allows us to see the vessels, much increased in size, arranged in the form of a close net-work upon the appendages of connective tissue, which separate the new or glandular part of the cotyledon into different compartments. The utricular glands and the follicles have also acquired a greater diameter; only they are perceived with more difficulty in proximity with the base of the glandular organ, and if we succeed in discovering them we find that they have lost the round form in their transverse diameter, and that they have acquired an elliptical form, or are even subdivided, and we are no longer able to distinguish the internal epithelial element, on account of their transparency.

In the glandular follicles I have sometimes been

able to see more easily the internal epithelium changed into very diaphanous granular globules (Plate VI., Fig. 1, *m, m*).

As to the structure of the new glandular organ, anatomists, ancient as well as modern, contented themselves with asserting that it presented externally several openings or points, which led Malpighi to compare it to the mushroom, called with us *sponzuola*. Panizza remarked, with somewhat more precision, that these external cavities are subdivided on the interior into several secondary alveoli. It is difficult, at the first glance, to form an exact idea of the internal anatomical arrangement of the glandular organ. I have tried to show it in diagram, Plate I., Fig. 2, and I hope that a precise idea may be had of the structure of this organ, by comparing it with Plate VI., Fig. 1, which represents in natural size, and also when magnified two hundred and fifty diameters, a portion of the glandular organ of a gravid cow, at the third month of gestation.

I have stated above that from the surface of the uterine cotyledon, in vertical sections, columns or pyramids of connective tissue are seen to rise (Plate VI., Fig. 1, *d, d*). They form the walls of corresponding calices, somewhat irregular, joined and crowded close together, which open outwardly through sluices or clefts of various form and extent, to allow a passage for the villi of the chorion. Over the internal surface of these calices are distributed, in very great numbers, lamellæ of connective tissue, which form tubes opening into the cavity. These tubes, covered over their inner surface and at their common orifice with pavement epithelium, constitute utricular glands, which are no longer vertical to the mucous membrane, as in the mare, but are placed along its transverse axis. Nor are they simple, as in the first case, but are superimposed upon each other. The form of the calices, the base being narrow and the outward opening quite large, renders it impossible to make vertical sections continuously upon any number of these utricular glands, as they are represented in the diagram. In fact, they are cut in every direction (Plate VI., Fig. 1, b, b, b, c, c). At the first glance one sees only a quantity of openings, without any order; then, on fixing the attention upon the section (b, b, b), at certain places, glandular follicles become visible, and, in their cavities, the cut chorial villi (a, a, c, c). There then remain for investigation the method of formation of the glandular organ, or of the portion of the cotyledon which is developed during pregnancy, and the retrograde metamorphoses by which the same organ disappears after delivery. I have been wholly without the materials necessary for making these observations.

Panizza, as I have said, had pointed out how the chorial villi, or cotyledons of the fœtus, were developed at the places corresponding to the uterine cotyledons, without stating anything precise upon the observation. Now I ought to note one fact that I have noticed in regard to the fœtal cotyledons.

I do not in the least doubt the assertion of Panizza; however, the examination of the chorion of cows in the third month of gestation, or a little earlier, deprived of the cellular envelope of its external surface which constitutes the decidua, and then subjected to the ordinary methods of imbibition, allowed me easily to establish the fact that the formative process of the cotyledons of the fœtus obtains largely over the whole external surface of the chorion, and that they are developed, in remarkable proportions, at those places only which correspond to the uterine cotyledons. Where these do not exist, the formative process of the fœtal cotyledons aborts, so to speak, and, with the aid of the microscope, is seen only in the form of certain small elevations, with broad, serpentine lines, formed of cells of fibrous tissues, which are supported upon a net-work of large corpuscles of connected tissue, with short and broad prolongations. On their walls we observe certain nuclei. The whole forms an extended and elegant mesh, which has lost the character of the corpuscles of the connective tissue without having acquired that of a real vascular network.

The poverty of resources at my disposal, and the great difficulties to be met with in this kind of research upon such expensive animals, have forced me entirely to neglect some other phenomena which I should very much have liked to make clear. If, in the future, favorable opportunities are still to be denied me, I hope they will be offered to others who will take advantage of them, in order to fill up the blanks which, in spite of myself, I am forced to leave in this work.

CHAPTER V.

ON THE GLANDULAR ORGAN, OR MATERNAL PORTION OF THE PLACENTA, IN ANIMALS WITH A SINGLE PLACENTA.

REALDO COLOMBO designated that part of the ovum which is brought into contact with the uterus of the mother, in the human species, by the name of placenta.[1] Accepted by the moderns, this name has been extended to all mammals, without regard to the form of the organ. However, in respect to difference of form, it is distinguished as disseminated or diffused, multiple, and single placenta.

Having already spoken of multiple and diffused or villous placenta, I am now to treat of single placenta, the structure and function of which have given rise to diversities of opinion that still prevail. I shall certainly not consider all of the many and various conceptions upon this subject, but shall simply confine myself to those that are in direct relation with the new observations I shall present, because I believe that even the imperfect observations and the errors

[1] Realdi Columbi de Re Anatomica, lib. xii., p. 248. Venice, 1559. Vesalius, before him, had called it orbicular flesh, and he was the first to give a special name to this part of the fœtal envelopes. At the present time it is generally believed that the term placenta is due to Fallopius, because in his Observationes Anatomicæ he wrote, "Carnem quæ placenta a me dicitur;" but Noortwyck has justly remarked in his book (Uter. Human. Fab. et Heist., page 116, Leyden, 1743) that the observations of Fallopius were not published until 1561, and consequently two years after the publication of the Anatomy of Realdo Colombo.

of our predecessors, in a determined order of ideas, are of great service in forming an exact judgment upon new researches.

In these historical studies in connection with the single placenta, it seems to me that the most important point to be explained and placed in the clearest light is the question whether the placenta belongs entirely to the fœtus, or whether it is distinguishable into two parts, one belonging to the mother and the other to the fœtus.

Harvey[1] had said concisely: "Placentam partem esse fœtus, non matris." This categorical formula, adopted by several authors, ought not to have been accepted, it would seem, after Hunter[2] had so clearly demonstrated the existence of the vessels which go from the uterus to the placenta, named by him utero-placental. The existence of these vessels was afterwards wrongly denied, and a considerable number of modern observers[3] have believed only in a simple superposition of the two organs; and instead of considering the decidua serotina as a means of union, they have attributed to it the characteristics of a thin and delicate inorganic tissue, destined to keep the two organs separate.

The idea of the placenta being separable into two parts, the maternal and the fœtal, is ancient, and, for certain animals at least, really belongs to Fabricius,

---

[1] Op. cit., p. 290.

[2] The Anatomy of the Human Gravid Uterus. London, 1794.

[3] Lee, Philos. Transact., 1832, page 57. Velpeau, Embry. ou Ovolog. humaine, Brussels, 1834, pages 63, 70. Radford, On the Structure of the Human Placenta, Manchester, 1832. Seiler, Die Gebärmutter und das Ei des Menschen, 1832, page 31. Ramsbotham, Millard, and Noble, London, Med. Gaz., 1834 and 1835.

who declared that in guinea-pigs the placenta is double.[1] However, the distinction of the two parts of the placenta is generally accorded to Wharton:[2] —

"Etenim ipsa placenta duplex est. Altera ejus medietas pertinet ad uterum, altera ad chorion. Atque hæ medietates inter se apte committuntur, seu potius inoculantur. Constat enim ex inæquali superficie: nimirum alveolis et protuberantiis sibi mutuo apte respondentibus: ita ut alveolis unius medietatis protuberantiam alterius in se excipiat et undique amplectantur."

The doctrine attributed to Wharton was soon combated, and Needham,[3] after reviewing the passage I have quoted, adds: —

"Hæc sententia ad literam vera est, si de ruminantibus sermo sit quibus omnibus uteri intima membrana statim a conceptu in tales glandulas exsurgit, et fœturæ preludat. At vero placentulæ a parte matris respondentes in solis glanduliferis [that is to say, the ruminants] occurrunt."

Notwithstanding this, Needham himself, who maintained that the placenta is simple, *et soli chorio propria*, in the carnivora as well as in woman, on comparing the gravid uterus in different animals, made with respect to the rodents an observation analogous to that which Fabricius had made upon the guinea-pig, and writes: —

"In cuniculo placentæ binæ sunt, neutra tamen uteri dicenda est, utpote quæ chorion in partu comi-

---

[1] "At in porculis indicis privatim, duplex carnea moles, altera alteri superposita observatur." (Op. cit., p. 39.)

[2] Adenographia sive Glandularum Totius Corporis Descriptio, page 218. Amsterdam, 1659.

[3] Op. cit., p. 26.

tantur cum eadem exeunt. Adeo ut haec animalia inter placentifera et glandulifera media videantur;" and a little further, still speaking of the placenta of rabbits:[1] "Singulis nempe fœtibus, singulas placentas impertit hoc animal, quæ tamen utero mediante glanduloso corpore agglutinantur."

This same fact was recorded by Graaf,[2] though opposed by him to the doctrine of Wharton:—

"In cuniculis vero et leporibus et quibusdam aliis animalibus ea parte, quâ chorio annectitur, rubet; altera vero, qua cum utero copulatur, albicat, et utraque cum fœtu simul excluditur, sic ut illa non magis quam altera ad uterum pertinere videatur."

Our own Malpighi, without describing them, divined more clearly than any other the two parts of the placenta, and assigned to each a different function. According to him, the placenta "est glandula conglobata sui generis, in qua portio uteri, propria carne donata, receptum ab uterinis arteriis succum percolat, qui separatus in sinuosis cavitatibus recolligitur, donec sensim fistulosas alterius glandulæ parti radiculas subeat, et venarum surculis excipiatur."[3]

The special observations and the doctrines of Fabricius, Needham, Graaf, and Malpighi were soon forgotten, and opposite statements were made.

Haller, for example, after having said that "cuniculi placenta ad humanum accedit,"[4] adds at page 243:—

"In cuniculo, placenta humanæ similis sanguine plenissima tuberculis suis ad similia tubercula uteri

---

[1] Op. cit., pp. 27, 189.
[2] Opera Omnia, page 207. Leyden, 1678.
[3] Opera Omnia, page 25. London.
[4] Element. Physiol., t. viii., p. 224. Berne, 1766.

adherescit;" and as to the glandular structure, he affirmed that there was no trace of glands in the placenta.[1]

Sixty years afterwards, Velpeau[2] declared that no one believed in the presence of glandular parts in the organ.

Many years did not pass, however, before the idea that Malpighi had formed of the placenta was revived by the observations of von Baer[3] and Sharpey and all those who believed that in some animals at least the utricular glands of the uterus entered into the structure of the placenta. At the same time they disguised, as it were, the fundamental idea, by supposing that the extreme ramifications of the foetal vessels of the villi of the chorion were brought into contact with the vascular net-work which envelops the uterine glands, and that by the extreme attenuation of their walls the exchange of materials between the blood of the mother and that of the foetus was always carried on in the same manner.

[1] "Ut nullæ veri nominis glandulæ in placenta sint." (Op. cit., p. 234.)
[2] Op. cit.
[3] The following are the thirty-first and the thirty-fourth of the general conclusions of von Baer's work, Zusätze zur Lehre vom Baue, und Verrichtungen der Geschlechtsorgane. Leipzig, 1846.

"It cannot yet be admitted that the villi of the chorion in woman penetrate into the uterine glands, as in dogs, since woman alone possessing a membrana decidua which distinguishes her from other mammals, there may exist a difference of formation between the human placenta and that of the dog.

"If, in the end, it should be demonstrated that the villi of the chorion in woman as well as in the dog penetrate into the uterine glands and distend them, it would follow that the ramifications and terminal filaments of the villi of the chorion take on a delicate covering which penetrates with them, and which they receive from the wall of the uterine glands. Even in that case we could hold the accepted doctrine upon the structure of the placenta and upon the mode of formation of its parts."

It must be confessed, however, that the observations of Sharpey and Weber gave rise to certain doubts in the mind of the illustrious Bischoff,[1] whose exact words it is advisable to reproduce: "If it be true," he wrote, "that the placenta of the dog, as Sharpey affirms, owes its origin to the penetration of the villi of the chorion into the glandular canals of the uterus, which are surrounded by a capillary network of uterine vessels, and that these canals and these villi, by continually augmenting in size and ramifying, become entangled with each other, as I have been able to prove by my observations upon the dog; if it be true also, as Weber and Sharpey affirm, that the human decidua is formed only in great measure by the highly developed uterine glands, and that its sieve-like appearance depends upon the openings of these glands, it becomes very probable that in the human species likewise the placenta owes its genesis to the fact that the chorial villi, constituting the umbilical vessels, penetrate into the glandular canals upon one point of the internal face of the uterus, and that both, continuing their development, finally constitute its organization. It is evident, then, that the meeting of the two bloods in the placenta does not consist in a direct interchange of materials between them, but that the vessels and the uterine glands elaborate a secretion, which is seized upon by the villi and the umbilical vessels that have penetrated into the glands."

The doubt expressed by Bischoff was welcomed by certain eminent physiologists, even in our own day.

[1] Entwickelungsgeschichte der Säugethiere und des Menschen. Leipsig, 1842.

No new fact has been brought forward, however, either to confirm or to contradict it. The opinion which has met with the greatest acceptation is that the uterine portion of the placenta is produced by the vessels of the mother which are conducted to it. Upon this point there was almost universal agreement. The divergences of opinion had reference to the manner in which the maternal vessels are brought into contact with those of the fœtus.

I have elsewhere said that I reckon among the defenders of the ancient doctrine, which maintained the direct communication of the blood between the mother and the fœtus, those of the moderns who, while rejecting the basis of the doctrine, claim that the exchange of materials for the nutrition of the fœtus is carried on by the process of endosmose and exosmose through the walls of the maternal and fœtal vessels.

The numerous defenders of this doctrine ought to be divided into two classes. In the first, I would rank all those who consider the placenta as nothing but a vascular net-work; the second would comprise those who, while accepting the principle, admit further that the vascular net-work of the maternal placenta is sustained by the folds of the mucous membrane of the uterus or by the decidua serotina in the human species.

Galen[1] was the leader of the first class: "Venas et arterias chorii in uteri venas et arterias insertas esse principium secundinarum," although he had said in another place[2] that the placenta is a glandular flesh which is formed around the vessels of the uterus.

[1] De Formato Fœtu.   [2] Aphorismi, 45.

Fabricius, among the ancients, maintained energetically the theory of direct vascular communication between the mother and the fœtus. Though combated by other observers, the opinion of Fabricius has always found, up to the present day, both friends[1] and opponents.[2] I have already quoted a few of them. I will add that the illustrious Italian anatomist, Panizza, summing up in his last work [3] the discussions that took place on the subject at the scientific congresses of Naples, Florence, and Padua, shows that several able men have been led into error by certain processes (injections), which seemed to prove direct vascular communication between the maternal and the fœtal blood; and he concludes that this direct vascular communication does not exist, but that the sanguineous system of the mother and that of the fœtus form a special circulatory apparatus, maintaining with each other an intimate and complicated vascular contact, even in the smallest loops of the arteries and veins.

Long before Panizza, Bischoff had acknowledged that there remained one problem to be solved: to determine how the maternal and the fœtal vascular sys-

---

[1] Cowper: The Anatomy of the Human Body, 1698. Noortwyck: Uteri Humani Gravidi Anatomia et Historia, 1743. Vieussens: Dissertatio de Structura et Usu Uteri et Placentæ Muliebris. Haller: Elementa Physiologiæ, t. viii., p. 255. Senac: Traité de la Structure du Cœur. (Paris, 1783.) Florens: Cours sur la Génération, page 138. (Paris, 1836.)

[2] Monro: Edinburgh Medical Essays, vol. ii., p. 68, 1749. William Hunter: The Anatom. Descript. of the Human Gravid Uterus. (London, 1794.) Wrisberg: Comment. Medic., 1800. Bischoff: Entwickelungsgeschichte der Säugethiere und des Menschen. (Leipzig, 1842.) Jacquemier: Archiv. génér., page 165. 1738.

[3] Sopra l'Utero gravido di Alcuni mammiferi, pages 14 and 15. Milan, 1866.

tems are disposed in the placenta, and what are the parts which serve to support these systems. It is generally believed that they tend more and more, from the beginning of their existence, to coalesce into one uniform organ, the placenta, in which the two parts, the maternal and the fœtal, are not distinguishable except at the first period of development.[1]

But as to the constitution, even primordial, of the maternal portion of the placenta, the only thing the best authors say about it is that "the vessels of the chorion are brought into relation with a corresponding and somewhat vascular portion of the uterine mucous membrane, of a reticulated and cellular appearance, and that this portion constitutes the uterine part of the placenta."[2]

Hirtl[3] makes the same observation, and writes that "upon one point of the uterine mucous membrane there is developed a colossal net-work of venous vessels, forming the so-called maternal placenta, which receives into itself the prolongations or projections of the embryonic placenta."

I pass over in silence the moderns who repeat the doctrine of the ancient anatomists, according to which the point of the uterus where the ovum is arrested becomes almost fungous (*carunculæ uterinæ*), and the fungosities, constituting the maternal portion of the

---

[1] Müller, Professor at the Veterinary School of Vienna, published at Berlin (Müller's Arch.) a memoir upon the structure of the placenta in the Dasyprocta Aguti. In this animal the two portions of the placenta remain distinct, though in vascular communication throughout the time of gestation.

[2] Bischoff, Op. cit., Dell' Allantoide.

[3] Manuale d'Anatomia topografica, t. ii., p. 112. Italian translation. Milan, 1858.

placenta, mingle with those of the chorion, or fœtal placenta, to form one single placenta. Madame Boivin and Velpeau,[1] who did not admit the double placental structure, said with reason: "If the placenta be thus formed, it remains to be known at what period and in what manner the fungosities or caruncles of the uterus are separated, the placenta being separated from the uterus by the decidua serotina."

The theory that each of the two placental parts is formed by very thin folds or sheets irregularly doubled over each other belongs to Eschricht.[2] According to him, those folds which constitute the fœtal placenta project vertically from the chorion. As to the maternal placenta, he declared that he had seen, especially in the cat, that the uterine portion forms a vascular membrane, which, while originating in the mucous membrane of the uterus, differs from it completely; it is transformed into very small folds, among which are set the villi of the chorion, which have also, he claims, a lamellated appearance.

Among the rodents, also, the placenta consists of a uniform crossing of the lamellæ of the chorion and of the vascular membrane of the uterus. Bischoff favored these opinions, and added that in rabbits, during the first period of gestation, when the placenta begins to develop, he had distinctly seen the folds of mucous membrane covered with an elegant vascular net-work. But in that animal he did not find uterine utricular glands inclosing the villi of the chorion, as is the case in the dog, in which latter animal he

---

[1] Embryologie et Ovologie humaine, page 46. Brussels, 1834.
[2] De Organis quæ Respirationi et Nutritioni Fœtus Mammalium inserviunt, pages 13 and 20. Copenhagen, 1837.

had been able to verify the observations of Sharpey and von Baer.

Müller[1] accepted these observations as demonstrated. He even set them up as a general principle, advancing the idea that the placenta presents two capital modifications: first, the development of branching villi of the chorion penetrating into the uterus; second, the formation in the uterus and the chorion of highly vascular folds which fit into each other. But shortly after, the same Müller,[2] in speaking of the human placenta, forgets his general teaching, and declares that that organ is composed of two elements or two portions, the fœtal and the uterine, which penetrate reciprocally into each other. The fœtal placenta consists of vascular trunks rich in branching villi; the uterine is formed of the substance of the decidua, which insinuates itself between the villi, and envelops them completely as far as the surface of the chorion. He adds that, according to Weber, the relation of these two parts is very different in the human being from what it is in the other mammalia. In the latter, the vascular villi of the fœtus are prolonged like roots into the equally vascular sheaths of the uterine placenta, so that the two systems of capillary vessels touch each other, and there is thus an exchange of materials between them. In the human species, on the contrary, the villi of the fœtal placenta penetrate deeply into the large sanguineous vessels proceeding from the maternal veins, so that the vascular loops of the fœtus are bathed in the blood of the mother.

[1] Manuel de Physiologie, t. ii., p. 730. Paris, 1851.
[2] Op. cit., p. 734.

On the other hand, Eschricht believed that in woman, as well as in the other mammals, it is only the capillary net-work of the decidua which enters into contact with the vascular loops of the fœtal villi. In this case, the structure of the uterine placenta would be in every way the same as that of the fœtal placenta.

In the midst of so many uncertainties, it gives me pleasure to call your attention to an observation of Cuvier, amplified by Müller,[1] namely, that in certain fishes (the *squalidæ*), such as the *carcharias*, the *prionodons*, and the *scoliodons*, the fœtus is united to the uterus by means of a placenta; and that, moreover, in those animals the uterine placenta is formed of very elevated folds of internal uterine membrane, which correspond exactly to the folds of the fœtal placenta, and they are as intimately blended as in the uterine and the fœtal placenta of the mammalia.

It has not been my privilege to confirm by my own researches this observation, which I do not hesitate to consider of very great importance for the class of subjects which I am laying before you. I pointed out at the beginning the change of the folds of the uterine mucous membrane into a glandular organ, when the placenta is single. The placenta of the *squalidæ*, according to Cuvier, would represent in the simplest and most elementary manner the placenta of the mammalia which have that organ single. Now, the perception of these progressive transitions of nature, as we may call them, always serves as a confirmative argument, and as an elucidation for the most complicated observations. It establishes the great

[1] Op. cit., p. 729.

secret of the high utility and of the ever new and ever fresh interest of studies in comparative anatomy. It is therefore with the greatest regret that I have been obliged to leave this blank unfilled.[1]

I shall now speak briefly of the principal opinions that have been admitted as to the functions of the placenta, further limiting my researches upon the subject to the main points which have or may have connection with the new observations which I shall present, putting aside altogether the functions, more or less probable, which observers have from time to time attributed to it, by induction.

The opinion of Fabricius, that the placenta served only to hold and preserve the vessels,[2] found no partisans. Haller wrote:[3] "Nemo, ut puto, negat, ab utero in fœtum, per placentam, succum aliquem alibilem percolari. De natura succi quæsitum est, quem uterus in placentam mittat."

Among the old writers, several conceived the idea that, even in the case where it is single, the placenta elaborates a kind of fluid, analogous to milk, to serve as nutrition for the fœtus. Although such a function was imagined without the support of any actual observation, yet the opinion of the ancients found advocates among certain of the moderns, who supposed the existence of such phenomena as would give it an appearance of truth. It is strange, however, that other real phenomena, having a certain value as support for the ancient doctrine, although

---

[1] See Appendix.

[2] "Præcipuum utilitatis scopum in hac efformanda fuisse vasorum custodiam atque propugnaculum." (Op. cit., p. 87.)

[3] Elementa Physiolog., t. viii., p. 238. Berne, 1766.

observed by able anatomists and physiologists, have not been sufficient to bring it forward again.

I believe that Wharton[1] was the first to affirm that the fluid elaborated by the placenta was in every respect like the milk of the mammalia. Harvey had merely contented himself with comparing, from a functional point of view, the placenta to the breast: "Jecur, inquam, est organum nutritium corporis, in quo est: mamma infantis; placenta embryonis."[2] Graaf was much more explicit: "Existimamus itaque," he wrote, "non sanguinem sed lacteum quemdam humorem esse, qui ab utero ad fœtum defertur."[3]

It would be tedious to name all those who accepted the doctrine of Graaf, modifying it somewhat,[4] or departing only slightly from it, and supposing the existence of special lacteal vessels destined to convey through the placenta the maternal chyle to the fœtus. The illustrious Haller, after having mentioned their opinions, does not hesitate to affirm: "Sed ii quidem viri[5] fabricam ruminantium animalium ad hominem traduxerunt."

Lauth[6] restored scientifically, so to speak, an opinion of Hippocrates by describing numerous lymphatic

---

[1] "Succus enim quem conficit in embryonis usum, lacteus plane est et lacti in mammilis genito simillimus." (Adenographia sive Glandularum Totius Corporis Descriptio. Amsterdam, 1659.)

[2] Exercitatio de Generatione Animalium, page 574. Padua, 1866.

[3] Opera Omnia, page 208. Leyden, 1678.

[4] "Aliquid forte lacti simile ad ovum ex utero venire." (Van-Swieten.)

[5] Verhejen, Vieussens, Falconet, Jenty, Deidier, Fizes, and others.

[6] "In muliere et pecoribus ejusmodi venulas et consimiles alias ad mammas et uteros ferri; quodque iis vehitur pingue, cum ad uterum pervenit lactis formam habere, ita ut puer, quod in sanguine dulcissimum est, ad sese attrahat simulque aliquantula lactis portionis fruatur." (De Natura Pueri, page 241. Geneva, 1657.)

filaments, of a special kind, which pass from the uterus to the placenta. Before Lauth, Everard[1] had described, in rabbits, chyliferous vessels going directly to the uterus. The observation of Everard received no more confirmation than that of Lauth. Finally, I should notice that Arantius had claimed that the placenta not only prepared, but purified, the blood for the fœtus; and since, in his time, this function was considered as belonging, for adults, to the liver, he named the placenta the uterine liver. Bartholin upheld this opinion. It is also this same opinion that Bernard[2] has restored to scientific repute in his recent observations, tending to demonstrate that there really does exist in the fœtus a placental hepatic organ, which produces glycogenic matter; an organ which disappears just in the same proportion as the liver of the fœtus performs the same function.

Prepossessed with such a conception, Bernard did not like to abandon it, even when he encountered glycogenic glandular cells in the placenta of certain animals. It was sufficient for him to have found sugar in the placental cells, to regard them as representing in themselves alone the glandular organ. I consider it useful, therefore, to give you at this point a summary of the observations of this illustrious physiologist; for although incomplete, they serve to confirm those I shall myself offer you. He wrote that[3] in the placenta of rabbits and guinea-pigs a whitish substance is met with, formed by agglomerated epithelial

[1] Novus et Genuinus Hominis Brutique Exortus, pages 132 and 282. Middelburg, 1661.
[2] Mémoire sur une nouvelle Fonction du Placenta. Paris, dans les Annales des Sciences naturelles. Série quatrième, t. x. Zoölogie.
[3] Op. cit., p. 113.

or glandular cells. He found these cells, like those of the liver of the adult animal, full of glycogenic matter, and they appeared to him to be placed in greatest numbers chiefly between the maternal portion and the fœtal portion of the placenta. It seemed to him also that these cells became atrophied in proportion as the fœtus approached the time of birth. Finally, he recognized that the placenta of rabbits and guinea-pigs is formed of two parts, which have distinct functions, — one part vascular and permanent until birth, the other part glandular, which prepares the glycogenic matter and which has a more limited duration.

In connection with these recent observations, we cannot forget the more ancient ones of Fabricius, Wharton, and Needham upon the placenta of the same animals. But, to return to Bernard and his researches, he proposed to himself to solve the question whether the function of the liver in the adult is likewise fulfilled by the hepatic placental organ, or whether, in the liver alone, the elements which elaborate the amylaceous matter and those which form the bile are distinct from each other.

I shall not enter upon this question, on account of the discussions which would arise, especially after the careful observations upon glycogenesis of my illustrious friend, Professor Schiff.[1] I shall note, however, one physiological and anatomical fact to be observed in the placenta of dogs. This fact, indicated by Marcus Aurelius Severin[2] and by Needham, and which I have myself confirmed, was neither observed nor inquired into by those of the ancients who, from the

---

[1] Nouvelles Recherches sur la Glycogénie animale. Paris, 1866.
[2] Zoötomia Democritea, page 307. Nuremberg, 1645.

point of view of function, compared the placenta to the liver. It was not even known by Bernard, who could have utilized it in his researches. In order to recall it, I shall quote the very words of Needham,[1] who, in describing the placenta of the dog, said, "In media parte tota rubet. Extremis vero lateribus utrinque viridis est: hujus rei ratio mihi nondum constat."

Neither have I been able to discover it. I will only add that the green matter in question is found more copiously elaborated by the epithelial cells covering the large folds of mucous membrane which rise around the placental edges, and the internal half of which has been transformed into a glandular organ in the placenta. We often find this green matter, which has the appearance of oil, accumulated between the folds or festoons of the mucous membrane. It dissolves completely in alcohol, imparting freely to the alcohol its own color. I believe that it would be worth the attention of chemists to determine the nature of it.

Those very authors who had seen the villi of the chorion penetrating into the utricular glands of the uterus did not imagine that the secretion of these glands is concerned in the nutrition of the fœtus. They either simply pointed out the fact, like Bischoff, in a doubtful manner, or else they thought that it might take place in certain animals only, as did Müller,[2] who made an excellent review of the observations of the most able modern anatomists and physiologists.

It would result from these observations that the

[1] Op. cit., p. 27.
[2] Manuel de Physiologie, page 736. Paris, 1867.

cellular arrangement which envelops the villi of the chorion during the period in which they are deprived of vessels is destined to take up the substances which the villi afterwards transmit to the embryonic cells; and that when the vessels have penetrated into these villi they take possession of the nutritive substance, in order to carry it to the fœtus, receiving it either from the maternal blood which in woman surrounds the villi, or from the white liquid of the uterine glands of animals. This meeting between the maternal fluids and the vessels of the villi in the fœtus also takes the place of respiration, and is its equivalent. Notwithstanding this, some modern physiologists of great reputation [1] acknowledge that they do not yet understand clearly how the exchange of materials is carried on between the mother and the fœtus in the mammalia.

In the mammalia we must distinguish two different periods of fœtal nutrition. At first, the ovum, while the placenta is not formed, finds the elements of its nutrition in the fluids which are poured out in abundance into the uterine cavity through the utricular glands; just as the young chicken finds in the white and in the yolk of the egg the albuminoid substances, the fat, the mineral matters, and the water necessary for its nutrition and development. But these elements not being sufficient for the complete development of the embryo, there is established always, though at different periods, according to the different classes of animals, an intimate bond of union between the mother and the fœtus. And though it has always been admitted by all authors that this

[1] Vierordt: Elementi della Fisiologia dell'Uomo, page 637. Milan, 1857.

relation is established by means of the placenta, we have just seen how many theories have been entertained as to the mode in which this organ performs its functions.

Although the opinions I have indicated have reference to a limited order of ideas, yet they leave a great uncertainty in the mind. This uncertainty would be greatly augmented if I were to quote the interpretations which were imagined without the aid of actual observation.

Now, I have had the good fortune to demonstrate, in a positive manner, the neoformation of a glandular organ over the whole internal surface of the uterus of an animal with a diffused or villous placenta, forming beyond all doubt its maternal portion. This fact, entirely new to anatomical science, aided me better to understand the same fact in cases where the placenta is multiple, as in the ruminants. These observations will also aid me now to demonstrate it in animals with a single placenta, in order to fix the idea firmly in our minds that always and under all circumstances, as I said at the outset, the relation between the mother and the fœtus is established by means of a glandular organ of new formation, and that this organ constitutes a really special or maternal part of the placenta, very different in structure and function from the other or fœtal part.

This fact once proved and this new truth assured to anatomical science, the physiological study of the fœtal nutrition and of the placental functions will find in clear and precise anatomical knowledge the basis which it lacked. It is not useless, therefore, to recall at this point the idea I have before advanced as to the existence or non-existence of the uterine mucous membrane in the human species and in certain animals. From the examination of the elements which compose it, I came to the conclusion that it always does exist; that its most simple and fundamental form is that of a more or less thick epithelial envelope; that its greatest complications depend on the elevation of the *sub-epithelial connective tissue;* and that on this elevation depend the different forms of simple or highly festooned folds, which, notwithstanding their development, always represent the extremely small epithelial follicles which we had observed upon the uterine mucous membrane of other animals. Now I will pass on to the demonstration of the fact I advance, and I hope thus to give a clear and exact idea of the formation of the maternal placenta, even in cases where that organ is single.

On the internal surface of the uterus of the mare, which is covered with the most simple and elementary mucous membrane, the new glandular organ, formed in pregnancy at term, represents, much enlarged, the small and delicate follicles which I described in the mucous membrane of the non-gravid uterus of the rabbit. Likewise, the new-formed glandular portion which is developed upon the permanent part of the uterine cotyledons of the cow, which are also covered, in the state of non-gestation, by an elementary mucous membrane, assumes more complicated and more perfect forms, analogous, however, to the types of those folds with multiple depressions which, like the simplest type, are observed upon the uterine mucous membrane of other animals in the state of non-gestation.

So far, then, the new glandular organ, or maternal portion of the placenta, follows in its character and in its development the same typical forms which we have observed in the uterine mucous membrane of various species of animals.

Now, on examining the same organ in animals with a single placenta, which possess normally a uterine mucous membrane with double folds or with the most elevated form of structure of this membrane, we shall see that, in these animals also, nature does not change the type of neoformation, and that it only modifies and amplifies preëxisting forms. In short, when the mucous follicles, or folds, or linings of the uterine mucous membrane preëxist, in certain animals, it is they which augment in number and volume during pregnancy, and which take on special exterior forms, without losing the typical form of a simple glandular follicle.

These facts sufficiently confirm what Leydig had supposed relative to the signification of the depressions or festoons of the folds of the mucous membrane in certain species of animals. This purely anatomical conclusion acquires even greater value, from an anatomical as well as from a physiological point of view, when applied to the study of the gravid uterus; for we are led to that other very important conclusion, that the new glandular organ developed during the pregnancy of animals is always the result of a transformation and perfecting of the uterine mucous membrane, without ever departing from the type which it presents in the normal condition of the non-gravid uterus.

The most simple forms of the placenta would be

those that Cuvier and Müller have described in certain fishes. Now these forms likewise would not depart from the general law established by my observations. They would rather demonstrate the same fact in its simplest and most elementary expression.

So far I have spoken of the single placenta in animals. Although all my predecessors, even the most illustrious, have sought aid and light for the study of the human placenta from the observation of that of animals, I believe that such a method must have injured the truth and become the sole cause of a great many errors that have been propagated. In the human species alone, the uterine mucous membrane never assumes, during pregnancy, the character of a more perfect membrane. Consequently, the new glandular organ which is developed differs greatly from that to be observed in animals. In woman, while preserving the fundamental and typical or characteristic parts of glandular organs, it loses completely all the accessory features or form-characters which we meet with in the different species of other mammals. I have said this from the beginning, and I repeat it at this point, that it may be understood why I am obliged to treat separately of the human placenta and that of animals, though the organ is single in both cases.

The placenta of animals is emphatically distinct from that of the human species, both in structure and in the disposition of the parts. The demonstration of this truth will be, if I mistake not, the greatest and most interesting conclusion of this work. I should consider myself happy if, in spite of the numerous gaps that I am forced to leave, I could at least fill up this one; and I should regard myself as more than happy, if it were granted me, at some future day, to produce the results of a comparative study between the placenta of the monkey and that of woman.

Do we belong, from the first moments of our existence, to the simian family, by the same placental structure, or, rather, does their placenta, like that of brutes, remove them from us, as we are already separated by the structure of that organ from all other mammals?

Is there among us any one who can answer this question, at once so important and so interesting? Good-will and courage are not sufficient for this kind of study; our poverty deprives us of all hope in this direction, for the love of science, which consoles us, is very often the saddest and keenest of misfortunes for us![1]

But before describing the structure of the fully-formed single placenta in animals, I believe it will not be without profit to present some observations upon the placenta in process of formation, obtained by examining the gravid uterus of a rabbit between the ninth and tenth days of gestation.

In speaking of the uterine glands, I have said that we find upon the mucous membrane of the rabbit numerous simple epithelial depressions, which I have called mucous follicles; and in speaking of the differences to be observed in the examination of the mucous membrane of different species of animals, I have called attention to the fact that the anatomical and physiological signification of these follicles is the

[1] See the last paragraph of the Appendix.

same as that of the numerous and often greatly elevated folds of this membrane, whether they be festooned or not.

In studying the formation of the placenta, I shall also adduce proof of this assertion. At present, it will be well to inquire what becomes of the very small mucous follicles in the gravid uterus of the rabbit.

In order that I may describe and explain, clearly as well as briefly, the observations made during this research, I make use of microscopic sections, represented at two hundred and fifty diameters, in Plate VII.

I ought to say, in the first place, that the uterine mucous membrane of the rabbit, between the ninth and tenth days of gestation, presents certain differences, according as we examine it near the place where the placenta began to be formed, or in the portions of the horns of the uterus that remain empty.

Near the placenta and in the portions of mucous membrane which, with the development of the fœtus, will be covered, or, more properly speaking, will be transformed into a placental organ, this membrane appears enlarged and tomentose. Examined with the microscope (Plate VII., Fig. 1), we see that this appearance is due to a remarkable development of those small follicles that I have indicated by the letters a, a. They are so close together that in the vertical sections of the figure we find several of them cut through at different points of their course (c, c). Around each of them we remark the proliferation of the connective tissue which forms the external wall of each follicle (d, d), and all around, as well as internally, the abundant formation of the epithelial element which is the essential part of all glandular organs. These same follicles, which we have seen so small in the non-gravid uterus, have acquired from 0.10 to 0.15 mm. in diameter, and from 0.22 to 0.30 mm. in length. Such hypertrophy at the place indicated is remarkable, for the outer edges of the growing placenta are made up of elevations or folds of the mucous membrane, which have on their exterior very apparent notches or festoons. In short, it is the outer edge of the last peripheral follicle, comprised in the formation of the placenta, which has been transformed into a festooned fringe of mucous membrane.

The vertical section of portions of the horns of the uterus that have remained empty allows us to see, instead of the simple development of the follicles, the mucous membrane rising and forming very prominent folds, which diminish more and more as they approach the enlargement of the horn where the ovum is arrested. On examining these folds with the microscope (Plate VII., Fig. 4), one is struck with the elegant figure presented. It seems difficult at first, to get a clear idea of the intimate structure, from the cut represented in the above-named figure; but on looking more attentively, we easily perceive that it is produced by cup-like or tunnel-shaped elevations of the mucous membrane, pressed close together and having an undulating course, coiling themselves towards their free surface. I shall soon return to the anatomical and physiological signification of these projections or complicated folds in the gravid uterus of the rabbit. The important point to be ascertained in the formation of the placenta is what has become of the follicles, so highly developed, which we saw in close proximity to the placenta at the beginning of its organization.

For this study, the transverse sections of the placenta have seemed to me most useful for the purpose of demonstration. Figure 2 of Plate VII. shows a section from the placenta of the rabbit in process of formation, near the uterine surface. The large openings, limited internally by epithelial cells (a, a), already indicate the remarkable calicular expansion which the follicles have speedily acquired at the spot where the placenta is formed. The placental base, consequently, has been formed by the development of the follicles, and by a very abundant proliferation of the sub-mucous connective tissue of the uterus (c, c). It is not uncommon to find in this part of the placenta, that some of the follicles, closely crowded against each other (b), have blended their walls together. In such cases, I have always observed that around the follicles in which the internal cavities, though contracted, were preserved, the transformation of the connective tissue into epithelial cells was more active throughout the whole interior of such cavity, — a fact which is not observed in the preceding case. In Figure 3 of Plate VII. I have represented another transverse section of the same placenta, but close to the fœtal surface. Here the cup-like cavities are seen rather narrower than at the base, and the form of already developed tubes predominates (a, a). I believe that this results from the sinuous and undulating course of the superior border of the calices which have grown out of the enlarged follicles, and which by frequent contact blend their walls together and form almost as many crowded and slender tubes which open upon the fœtal surface of the placenta.[1]

Indeed, from vertical sections of this part of the placenta in process of formation, we see by the microscope that the fœtal surface is entirely covered with tracings, and that it has a tomentose appearance like a sponge. At the aforesaid period of pregnancy in the rabbit, we already perceive the openings of the tubes or canals which are nearly formed, and are filled with small oval cells, colorless and very transparent. When the vascular villi of the chorion are completely developed, they will find the openings of the glandular organ ready to receive them.

Between the fœtal surface of the placenta in process of formation and the neighboring part which 1 have had drawn in Figure 3 of Plate VII., there is not only the difference that I have indicated in relation to the great number of orifices of the glandular follicles, crowded close together, but there is the further difference of a smaller quantity of true connective tissue interposed, having here rather the characteristics of a gelatinous substance than of an aggregation of corpuscles of connective tissue. Hence the spongy and tomentose form noticed above. When the vascular villi of the chorion are formed and have penetrated into this gelatinous tissue, there are organized large round cells, analogous to those of which we shall see the membrana serotina composed in the human species. They are neutral cells, which, by their multiple and rapid transformations, serve to perfect and complete the structure of the normal pla-

[1] See Appendix.

centa and to establish an intimate relation with the utero-placental connective tissue, by means of which it is blended with that of the chorion.

We shall see presently, more clearly, that these same rudimentary parts are preserved, with the exception of some difference of form, in the completely organized placenta. But I wish first to direct your attention to the projections or folds of the mucous membrane which I have described in the unoccupied portions of the uterine horns; for it appears to me that they give better than any description a precise idea of the internal skeleton, so to speak, of the glandular organ or maternal placenta of animals with a single placenta. In a word, these folds of the mucous membrane (Plate VII., Fig. 4) are the rudimentary portions of the glandular organ or maternal part of the placenta, which has been arrested in its development, solely because the fecundated ovum has fixed itself at another point. This fact offers another quite important consideration for the physiologist, namely, that at the rutting period, and, still better, after fecundation, the whole internal surface of the uterine horns is prepared to be transformed into a placenta. This ought naturally to be the case, the point of the uterus where the ovum is to fix itself not being determined by any law.

In the parts which form the glandular or maternal portion of the placenta, there take place successively other important transformations, which are established when the vessels appear in the fœtal villi of the chorion, and when the neutral cells above referred to are formed. I do not insist upon this point, which is not yet sufficiently studied, and which, consequently, the surface of the chorion (Plate VIII., Fig. 1, *c, c*), these tubes terminate in openings of variable diameter, according to the number of tubes that are confluent into one opening. This fact, which is easily observed by using vertical sections from the fœtal surface of the placenta of the dog, has not been remarked by any one, unless Haller may have had reference to it when he wrote: "In cane dum villosum chorion, foraminulentum et reticulatum, detrahebam, succus serosus defluxit, et alius sanguinis successit."[1] In any case, I have satisfaction in quoting these words of the illustrious physiologist; for though brief and ambiguous, they have, I think, an important bearing upon the observations that I have just presented to you, notwithstanding the utter neglect in which they have been left. The villi of the chorion penetrate into the orifices of the above-named sinuous tubes.

It now remains for us to ask what the relation is between these tubes and the glandular organs formed by the enlargement of the depressions between the folds of mucous membrane which, as I have pointed out, are seen upon the uterine surface of the placenta by both vertical and transverse sections (Plate IX., *d, d*).

Vertical sections are very useful for this purpose. At the bottom of Plate IX., *a, a*, is represented the connective tissue of the uterus, which, having become elevated by proliferation, has lifted higher the large festooned folds of mucous membrane, whose extremity or *cul-de-sac* is represented at *d, d*, with magnified volume.[2] The walls of these festoons or follicles,

---

[1] Op. cit., t. viii., p. 243.   [2] See Appendix.

already transformed into glandular organs, become blended together and go to make up larger cavities, surrounded by a peculiar layer of epithelium, which, as it is detached, forms a milky fluid that coagulates in alcohol. This fluid fills certain parts of the cavity into which it is poured. Upon the superior walls of these cavities open the tubes of which I have spoken (e, e); or, to speak more properly, these walls are continuous with those which form the external wall of the tubes, while the internal cellular wall of the cavity is continuous with their internal wall. The villi of the chorion, the fœtal placenta, do not penetrate into the inferior glandular cavities of the placenta, formed by the folds of the mucous membrane.

These observations probably explain how Eschricht and Bischoff may have confounded these glandular cavities with the dilatations of the utricular glands that I have represented at b, b, and how from the sinuous progress of the villi of the chorion within the winding canals of the glandular tubes may have also arisen the idea of laminæ constituting the fœtal placenta of the uterus, crossing and meeting each other in different ways (Plate VIII., Fig. 2). The fact is that, even in the case of single placenta in animals, it is the mucous membrane that is transformed into a glandular organ; and this organ does not lose the type of simple glandular follicle, merely because one part of the follicles has a singularly flexuous arrangement and because they communicate with each other (Plate VIII., Figs. 1 and 2, e, e). We always find the form of a canal or tube, at one extremity of which is the orifice, the other terminating in a *cul-de-sac*, in which is inclosed a vascular loop of the chorion, charged with absorbing the fluid resulting from the internal epithelial product of the glandular follicle.[1]

Even in the cases of single placenta, the maternal portion of it is glandular and the fœtal vascular. The relation between these parts is more complete than where the placenta is multiple or disseminated, but there is very little difficulty in referring the structure back to its most simple and elementary forms. To convince ourselves of this, it is only necessary to examine Figure 1 of Plate IV., representing the structure of the placenta of the mare, and compare it with those which have served for the description of the placenta of the dog.

The same structure is observed in the placenta of the cat; but for the descriptions as well as for the drawings, I have preferred that of the dog, because the greater volume of the parts in this animal renders observation more easy.

Now, on bringing together these observations, drawn from researches made upon the placenta during its development, the connective link which unites the different parts described is obvious. The base of the placenta in process of formation (Plate VII., Fig. 2) corresponds entirely to what I have described and drawn in Plate IX., *d, d*. The glandular tubes, already formed, represented in Plate VIII., Fig. 2, correspond to those that I have had drawn in process of formation in Plate VII., Fig. 3. Finally, the superficial, perforated, and spongy layer that I have described in the growing placenta is found indicated

---

[1] For greater clearness compare the diagram of the placenta of the dog (Plate I., Fig. 3) with the microscopic sections described and represented (Plates VIII. and IX).

in the fully developed placenta represented in Plate VIII., Fig. 2.

It is generally thought that in the delivery of these animals the fœtal as well as the maternal portion of the placenta is expelled from the uterus, and that there is consequently a true traumatic lesion; and for this reason Weber[1] proposed to divide the mammalia into two classes, according to the mode of separation between the two parts of the placenta, at the moment of parturition. In the first class he comprised all those animals in which the placentæ are united so slightly that they separate without producing in the uterus the slightest traumatic lesion, as in the ruminants. In that case, according to him, the maternal placenta remains in the uterus, simply decreasing in volume. In the second class he placed the animals in which the two placentæ are so clearly united that, at the moment of parturition, the uterine is detached at the same time as the fœtal, so as always to produce a traumatic lesion of the uterus. Among the latter he classed the carnivora, the rodents, and woman. For them all, the single placenta is only a caducous organ, which must be reproduced at every pregnancy.

Such a view of the subject, though adopted by the most learned anatomists and physiologists, is utterly incorrect. We have seen, beyond the possibility of doubt, how, in the solipeds, the maternal portion of the placenta, which is disseminated or diffused over the whole internal uterine surface, not only diminishes in volume, but is entirely destroyed, and disap-

---

[1] Froriep's Notizen aus dem Gebiete der Natur und Heilkunde, No. 996. 1855.

pears after delivery; and how the same thing occurs to the portions of the cotyledons which are developed during the pregnancy of those animals in which the cotyledons of the non-gravid uterus indicate only the points where the neoformation of the maternal placenta will take place during pregnancy. It follows from these observations that in all cases, without exception, the maternal placenta is a caducous organ at the moment of parturition or afterwards, and that it must be reproduced at every pregnancy. As to the traumatic lesion of the uterus, admitted by Weber for the carnivora as well as for woman, I must repeat that such an assertion is also entirely incorrect.

Examining, indeed, the uterus of dogs at different periods after delivery, we observe certain things worthy of attention. In the uterus of a dog killed two hours after parturition, I found the cavity noticeably stained by a fluid or mucus of a dark green color. This coloration is produced by the green matter elaborated in great quantity, as I have said before, by the epithelial cells of the great folds of mucous membrane which surrounds the placenta. And what becomes still more important is that it is these very elevated folds that remain in the uterus and circumscribe the place where the placenta was formed, and this place, instead of being denuded by the loosening of the placenta, is found covered, on the contrary, with a layer which, from its external appearance, one would take for a thickened mucous membrane, in continuous relation with the folds which limit the placental region.

This is sufficient to show that traumatic lesion of the uterus in animals with a single placenta is very slight and very limited. The appearances are different if we examine the uterus of those animals two or or even six days after delivery. Then the green substance has disappeared; and, in the contraction of the uterus returning to its normal volume, the thickened mucous membrane which I remarked at the places where the placenta had been formed is found changed into numerous folds, close together, much elevated, and rather thin. What the membrane of which they are formed has lost in thickness, they have gained in length. The folding into close-lying fringes is the effect of the remarkable diminution in the size of the uterus.

I was not able to repeat the observation until thirty days after delivery. The place where the placenta had been formed was still indicated by a thin, blackish crust, about as large as a penny. The animal which had served for the examination was of medium size. It is then demonstrated that the large folds of uterine mucous membrane which rise at the edges of the placenta and contribute to its function with their innermost portion, as well as the strong layer, like thickened mucous membrane, which occupies the whole place where the placentæ were organized, gradually change and disappear, in order to make way for the production of the mucous membrane proper to the non-gravid uterus. Briefly, in delivery it is the new portion of mucous membrane transformed into a placenta which is detached. The old mucous membrane remains, in a thickened condition, undergoing as a consequence of the uterine contractions the modifications indicated, to be at length slowly destroyed altogether.

Now, on observing the large folds on the second and even on the tenth day after parturition, it has seemed to me that their progressive destruction is accomplished by means of a special and limited fatty degeneration. In this observation, as in the others which I have placed before you, I am forced to leave many deficiencies, on account of the serious difficulties to be encountered at each step in this kind of investigation. I have not liked to pass over in silence the little I have been able to do, because, with the paucity of observations that we have on this subject, the facts stated may, in some degree, aid us to comprehend how occurs after delivery the destruction of the glandular organ over the whole uterine surface of the mare and in the glandular portion of new formation of the uterine cotyledons of the cow, and may greatly modify the opinions at present entertained on the subject.

Even from Weber's point of view, man cannot be confounded with the animals. The traumatic lesion of the uterus, in consequence of delivery, either does not exist, as in cases of disseminated or multiple placenta, or it is very slight, when the placenta is single, as I have demonstrated in the dog. In woman only delivery produces a true and considerable traumatic lesion of the uterus. This is what I am about to demonstrate, in passing to the description of the structure of the human placenta.

CHAPTER VI.

THE HUMAN PLACENTA.

I HAVE several times stated that the structure of the human placenta differs greatly from that of the mammiferous animals, even those having a single placenta, not only in the shape and the disposition of its various parts, but in the manner of its formation and of its separation from the uterus in delivery. Notwithstanding these remarkable and important differences, there exists also in the human placenta, as in the placenta of all mammals, the glandular organ of new formation, constituting the maternal portion, which is brought into direct relation with the villi of the chorion, that is to say, the fœtal placenta.

The real cause of all these differences is to be sought in the phenomena which are produced at the surface of the uterus upon which the placenta is formed, in order to give rise to the neoformation of the glandular organ; in the manner in which the villi of the chorion are brought into relation with this same organ; and, finally, in the mode of distribution of the utero-placental vessels in the human placenta.

The most notable modification that takes place upon the uterine surface, at the point where the placenta appears, is connected with the formation of the membrane called decidua serotina, a membrane which is lacking in animals.[1] In the latter, the development of the maternal or glandular part depends upon a direct metamorphosis of the uterine mucous membrane. But before inquiring into the intimate structure and functions of the membrane in question, it will be proper to quote in the first place a few of the numerous opinions that have been held, both as to the mode of formation and as to the structure of the true or uterine decidua. I shall speak particularly of those only which can aid me to demonstrate the very great importance that the decidua serotina has in the formation of the placenta and for the nutrition of the human fœtus. It is this point, I think, which deserves the most attention.

As to the mode of development and the relations of the decidua to the ovum, we encounter wide diversities of opinion. According to the observations of the most modern physiologists, of Vierordt, for example, the ovum, after its entrance into the uterine cavity, adheres to some point of the mucous membrane; the membrane then swells around the ovum, embraces it like a capsule, and makes a nest for it, as it were, or constitutes, as it is named, the internal or reflected decidua. This mucous capsule enlarges along with the ovum more and more, and at the third month it becomes attached to the uterine mucous membrane which is covered with the external decidua;[2] blended with the latter, it then makes but one single membrane, which gradually loses its vascularity.

Formerly, it was believed that there was an entirely

---

[1] See Appendix.

[2] Vierordt would prefer the terms external and internal decidua to those generally employed, uterine or true decidua and reflected decidua.

different process. It was supposed that the uterine orifices of the Fallopian tubes became obstructed by the tumefaction of the mucous membrane, and that this membrane then enveloped the ovum at its entrance into the uterus. Consequently, that part of the mucous membrane which, pushed by the ovum in its passage from the tube to the uterus, was folded back upon itself, and thus enveloped the ovum directly, was distinguished as the reflected decidua. This was the mucous capsule of Vierordt. The true decidua was formed by the portion of the uterine mucous membrane not folded back, which thus represented the external layer of the envelopes of the ovum.

The principal opinions held afterwards as to the structure of the membrana decidua in the human species are reduced to the following: it is either anhistous, analogous to the products of exudation or to pseudo-membranes; or else it is a special, organized membrane. Among the partisans of the latter, we must distinguish those who believe that the decidua is formed by a sort of tumefaction or expansion of the uterine mucous membrane ("membrana uteri interna evoluta," as Seiler says), granting to it an organization and vessels, from those who, after the teachings of Weber, claim, with Bischoff, that the decidua is made up from the layer of the internal glands of the uterus, in the midst of which are found numerous vessels with an exuded, plastic matter, which is so intimately united to the internal uterine mucous membrane that it seems to form one and the same structure. This idea appeared to be well supported by the observation of the fact that by compressing the walls of a gravid uterus we see the fluid burst forth from the utricular glands and from the numerous orifices with which the surface of the decidua is pierced, when it is completely detached from the uterus. These facts seemed to show the continuity of the glandular canals with the decidua.

In speaking of the uterine glands, I have noted how in the decidua of the ruminants, which has the form of a thin epithelial envelope, we observe certain small opercula, found upon the external surface of the chorion, and which correspond to the orifices of these uterine glands. In the human species, on the contrary, the epithelium of the mucous membrane and that of the uterine glands mingled with the organizable substance constitute the thick membrane called decidua; the latter adheres to the internal epithelial surface of the uterus, and the glandular secretion is so active and so continuous that it keeps the orifices of the glands open through the decidua. These numerous openings in the decidua give to it, when it is detached, the appearance of a membrane pierced like a sieve.

During menstruation, there is also organized a species of decidua, sometimes quite remarkable, named catamenial decidua, which is always highly developed after conception.[1] Such facts led Hunter, Seiler, Shar-

---

[1] Quite recently the changes which the mucous membrane of the uterus undergoes during menstruation have been carefully studied by a number of good observers. Dr. J. H. Aveling published an article in the Obstetrical Journal of Great Britain and Ireland in July, 1874, upon Nidation in the Human Female, which process he defines to be the periodical formation of the membrane lining the body of the uterus, which is developed during the inter-menstrual period, and is dependent upon ovulation. He states that, the nidal decidua having reached its full development,

pey, and Weber to affirm that the uterine mucous membrane is already prepared during menstruation and no impregnated ovum having arrived to demand from it protection and sustenance, a process of degeneration takes place, its attachments are loosened, and it is expelled by the contractions of the uterus, sometimes whole, in the shape of a triangular sac, but more frequently in minute portions. How long this process occupies has not been determined, but it is probably completed during the menstrual period. The act of denidation probably determines that of menstruation, because it is from the denuded surface of the uterus, caused by the removal of the nidal decidua, that the menstrual flow comes.

Dr. John Williams published a very interesting article in the same journal for March, 1875, in which he claims that there is no interval of uterine rest, but that the nearest approach to this condition is during the flowing, when the mucous membrane is undergoing disintegration. Even then the subjacent muscular wall is in a state of active preparation for the formation of a new mucous membrane. Menstruation, according to him, is produced by a process of fatty degeneration of the uterine mucous membrane, always commencing just within the internal os, and gradually extending to the fundus. The time necessary for the removal of the membrane varies in different individuals. Disintegration not only extends to the glandular and mucous elements, but includes the walls of the superficial vessels, so that a lesion occurs and hæmorrhage takes place. The development of the new mucous membrane follows the destruction of the old *pari passu*, and is completed in about a week. This, he declares, is formed from the uterine wall direct, "muscular fibres producing the fusiform cells, the connective tissue the round cells, and the groups of round cells in the meshes formed by the muscular bundles the granular epithelium."

Dr. George J. Engelmann, of St. Louis, in an article published in the American Journal of Obstetrics, May, 1875, denies that processes occur such as Aveling and Williams describe. From carefully repeated microscopic studies he shows that the mucous membrane during menstruation is swollen; that its surface is puffy and wavy, and after removing the mucus it becomes more distinctly visible; and in vertical sections the glands themselves may be seen without the aid of a lens in their entire length as white striæ.

The microscopic changes of the membrane are equally noteworthy, showing a remarkable proliferation of cell character, producing hypertrophy of its superficial layers, and raising its surface about the ostia of the glands, so that they appear like funnel-shaped depressions. The glands, except at their fundi, are very much enlarged, often two and four fold.

"The facts gathered warrant the conclusion that the mucous membrane to receive the ovum, and that after conception it is transformed into the external envelope of the ovum.

Now if, according to the preceding observations, the uterine or true decidua is only a product of exudation, due in great measure to the secretion of the uterine glands and to the transudation of organizable fluids from the internal surface of the uterus, it becomes easy to comprehend the formation of the reflected decidua which has been the object of so many discussions among anatomists. While the whole internal surface of the uterus is covered with the uterine decidua, the ovum when reaching the uterus is also covered with a similar layer, formed by the same materials which further serve to fix it at some one point of the uterus. At this point the uterine and the reflected deciduæ are soon mingled together, precisely as will take place later, when the ovum, increasing in size, forces the reflected membrane which surrounds it against the uterine decidua, so that the two deciduæ blend into one single membrane.

The uterine and the reflected deciduæ are composed then of the organic materials elaborated by the internal surface of the uterus, and by the uterine glands. These products serve to nourish the ovum

of the womb begins to increase in thickness as the time of menstruation approaches; that this tumefaction is most marked during the period itself, and gradually decreases after the cessation of the catamenial discharge. The hæmorrhage in the menstrual womb is always confined to the surface of the lining membrane, and the fatty degeneration is likewise more marked in its upper layer."

My own studies confirm me in the belief that the conclusions of Dr. Engelmann are correct, and that the uterine mucosa does not depart in its function physiologically from that of other mucous membranes in such an extraordinary manner as to destroy itself and be regenerated at every normal menstruation. — H. O. M.

before the formation of the glandular organ, with which the villi of the fœtal placenta will be brought into relation when they shall have become vascular. This easily explains the fusion of the two deciduæ. The uterine decidua is probably thus developed in the human species, because the uterine mucous membrane does not rise in numerous and distinct folds, constituting enormous open glandular follicles, as it does in animals; it is by reason of this that in the latter a species of rudimentary placenta is formed from the first moments of conception, and perhaps from the time even that the rutting season begins. But if all this is not difficult to understand, and is confirmed by more or less exact observations of illustrious authors, the same cannot be said for the decidua serotina, which, according to some, is simply the place or point where the two deciduæ have come in contact with each other, to fix the ovum against the wall of the uterus.

The ancient anatomists remarked that the uterine surface of the human placenta is rendered uneven by the presence of elevations or rounded lobules, crowded together. Arantius[1] taught that this placental surface, called by him fungous, was covered with a delicate, thin, and whitish membrane. Fallopius[2] believed this membrane to be formed by an agglutinant matter. Bojanus afterwards called it membrana serotina, and others utero-placental plate.

No modern author denies its existence; but no agreement has been arrived at as to its formation and function. It has been called a subtile lamina of

[1] Arantius: De Humano Fœtu, cap. 10, p. 71.
[2] Observ. Anatom. in Operibus Vesalei, page 754.

the uterine decidua, having the same structure as the latter. Some attribute vessels to it; others dispute their existence. It has been claimed that it served to unite the placental lobes or cotyledons, and the idea has been sustained that it penetrated not only between the cotyledons, but also between all the vascular filaments. Some have supposed that it is to be met with in all the periods of gestation; others, that it exists only during the last three or four months. According to Haller, Rouhault[1] was the first to observe that the serotina extends as far as the concave or fœtal surface of the placenta, through the clefts to be seen upon its uterine or convex surface. Hobokenius noted that the chorion, which covers the placental furrows and descends into them,[2] serves to unite the lobes of the placenta with each other.[3] Thus we find that the chorion and the decidua serotina have been confounded.[4]

About thirty years ago, Burns[5] maintained that the placenta was the result of the mingling of the uterine vessels with the external layer of the membrana decidua.

Müller[6] claims, that the uterine placenta of woman is formed by the development of the decidua,

[1] Rouhault denied, without reason, a membranous structure to the serotina, because it is pierced with numerous holes which allowed passage, as he thought, to a great number of vessels. (Mém. de l'Ac. R. des Sciences, page 182. 1714.)

[2] Anatomia Secundinæ Humanæ Repetita, page 113. Utrecht, 1675.

[3] Noortwyk believed for some time that the observation of Hobokenius was exact, but he afterwards denied it altogether. (Op. cit., p. 153.)

[4] "Frequentissime etiam a partu; latas lacinias chorii reperio quae in utero, manscrint." (Haller, op. cit., t. viii., p. 235.)

[5] Medical Gazette of London.

[6] Op. cit., t. ii., p. 731.

which enlarges towards the fœtal placenta and insinuates itself between the bundles of its villi, so as to join the internal surface of the chorion. He adds that the structure of the decidua serotina or secondary decidua is analogous to that of the true decidua.[1]

Vierordt does not depart very far from these assertions when he writes that "at the part of the chorion which faces or touches the wall of the uterus, the villi are developed; and that upon the uterine surface are formed the corresponding portion of the decidua, and the sanguineous vessels of the mother. From the fibro-muscular layer of the uterus," says he, "depart numerous small arteries which are distributed to this portion of the decidua, there branching off into capillaries; then their walls become more and more attenuated, and finally they disappear. During this period, the part of the decidua which has been often called decidua serotina is found riddled by areoli or cavities which contain blood, and which develop to such a degree that there remains almost nothing of its primitive and fundamental conformation."[2]

Thus it appears that the uncertainties which prevail in science as to the decidua serotina are very great.

In order not to have dwelt so long upon the subject to no purpose, I shall now explain my general idea concerning this membrane, as I have already done upon the uterine and the reflected deciduæ. I

---

[1] Op. cit., t. ii., p. 714.
[2] Fisiologia dell' Uomo, page 789. Milan, 1862.

have said that the place where the ovum is arrested is also the point where, in the beginning, the two deciduæ come in contact and are blended, however thin the reflected membrane which surrounds the ovum may be at that time. Every one knows that it is here particularly that the villi of the chorion are developed, which are at first the only means by which the ovum absorbs the nutritive elements from the materials that surround it. Not only do these villi increase greatly in magnitude, but they also become more complete, a fact due to the vessels proceeding from the allantois which traverse them.

Absorption ought to be, consequently, very active at this point; and it is proved that this is the fact by the considerable development of the villi. It follows that here especially the component elements of the deciduæ are more rapidly absorbed than elsewhere. During this time, the connective tissue of the subjacent internal uterine surface proliferates, and is transformed to such a degree as to constitute the most marvelous tissue known in the animal organism.

The structure of the decidua serotina is, as we shall observe, so peculiar as to prevent its being confounded with the uterine and the reflected deciduæ, a confusion which has been almost universal to the present day. We shall further see in the sequel, that in addition to very remarkable peculiarities of structure, the serotina ought to be considered as the stroma of the glandular organ or maternal portion of the placenta. For such is the noble and important function which nature has intrusted to this membrane, to which anatomists and physiologists have always attached so little importance; I shall continue to call it the serotina so as not to cause confusion by introducing a new nomenclature.

Before giving you my own observations, I ought to fulfill the duty of historian. I shall do so with all the more pleasure because in quoting the microscopic observations of Robin [1] and of Pouchet [2] upon the structure of the decidua serotina, it will be necessary to give the earliest researches that were made upon this subject, and that have not yet been repeated by other observers. Though incomplete, they support those that I shall myself offer you.

Robin regards the grayish membrane which is seen upon the uterine surface of the placenta, and which is buried between the cotyledons, there augmenting in volume, as a portion of the decidua. This membrane, he thinks, is visibly continuous with the portion of the decidua which adheres to the chorion. The grayish tissue which forms it contains: first, an amorphous matter; secondly, molecular granulations of different nature; thirdly, cells which have undergone a considerable hypertrophy and which present the most diverse changes, simulating all the known and described varieties of cancerous cells. To discover these cells, which have all the characteristic elements of cancer, it is only necessary, he adds, to scrape the uterine surface of a fresh placenta with the blade of the scalpel, and to examine with the microscope the pulp thus obtained.

These epithelia so entirely altered are, however,

---

[1] Mémoire sur quelques Points de l'Anatomie et de la Physiologie de la muqueuse Utérine. Paris, 1858.

[2] Précis d'Histologie humaine, page 359. Paris, 1864.

only normal elements or cells of the uterine mucous membrane that have undergone, in the midst of a normal tissue, physiological modifications which such elements ordinarily present, in the animal economy, only under a pathological influence. The nutrition of the fœtus is carried on by means of an endosmotic interchange through this grayish layer which constitutes the decidua serotina.

The results of the observations of Robin [1] are exact only in part, namely, in what relates to the existence of very singular cells in the decidua serotina. All the rest is erroneous, as I am about to demonstrate. I have stated that the decidua serotina is distinguished from the true or uterine decidua and from the reflected decidua by anatomical characters. In order to prove what I have advanced, I shall begin with the study of the placental villi, because in investigating the intimate structure of these and their relation to the decidua serotina, I shall necessarily describe the structure of the latter, and so better explain its functions.

The ancients were aware that the placental villi, which they called branching fibres, were surrounded with something that most authors were content to call a cellular web. Hobokenius spoke of it as nervous substance; Needham said that it was gelatine: "In gelatinam leutam et multis in locis glandulosam concrescit ut difficilius a vasis separari possit." [2] Stuart called it a special carneo-spongy substance. Haller considered that " cum truncis vasorum advenit, qui-

---

[1] Dictionnaire de Médecine, etc. Tenth edition. Paris, 1858. Art. Placenta.

[2] Op. cit., p. 34.

bus est pro vagina, indeque ad minimas usque fibrillas comitatur;" and according to Albinus, "altera superest tenerior cellulositas quæ a chorio propagata vascula singula obvolvit."

Thus he concluded that the placenta was composed only of vessels and of the cellular web. As to the function of this web, he contented himself with affirming that there were " clarissimi viri, qui ex utero sanguinem malunt in cellulosam tetam deponi indeque per venulas placentæ resorberi." [1]

In the works of the moderns we find that it is a denser, more fragile, and less regular layer of the utero-placental plate; and that it surrounds the vascular trunks. This led some among them to believe that the placental vessels were diffused throughout the thickness of the decidua itself, or that the chorion was composed of several sheets or leaves; that the decidua or anhistous membrane sent one leaf to the external surface, and another to the fœtal surface of the placenta; and that the delicate pellicle formed by the latter was folded between all the lobes, lobules, and vessels of the placenta.

We vainly seek in recent works for fuller information upon this subject. We must except, however, that given by Farre,[2] which it is well to present, as I have not seen his opinion upon this matter quoted by any one. According to him, each villus is composed of two distinct parts; that is to say, of an external membranous layer, and of a softer and vascular internal tissue, which is introduced into the other

---

[1] Op. cit., p. 241.

[2] Farre, Arthur: the article, Uterus and its Appendages, from the Cyclopedia of Anatomy and Physiology, page 718, fig. 485. London, 1858.

like the fingers in a glove. The distinction between these two tissues is not easily observed, except in cases where the external layer is torn, leaving the internal substance exposed, or in case it happens that the placenta remaining for some days in the uterus and undergoing changes there, its internal part shrivels and separates from the external wall, so as to leave a small space between them. On examining a villus microscopically, we see that the external layer is formed by a transparent membrane, amorphous, and non-vascular, in which are set, or attached to the internal surface, numerous spheroidal cells, slightly flattened, forming a single layer.

The internal portion, according to Farre, is formed of a soft and pulpy tissue which envelops the blood-vessels of the villi, and in the midst of which are found inclosed other numerous cells of the same nature as those which exist upon the internal surface of the external layer.

About ten years afterwards Vierordt, without being aware of the observations of Farre, confined himself to teaching that the villi are formed by a vascular loop of connective tissue, and covered with an epithelium which would favor, he thought, the endosmotic exchange with the blood of the mother.

But Farre does not inquire into the origin of the external layer of the villi which he had in part so well described; and even while saying [1] that the villi come from the chorion and extend to the decidua, in which they become implanted and adhere as far as they penetrate into it, he excludes what is really observed in the phenomenon, namely, that it is the de-

---

[1] Loc. cit., p. 719.

cidua serotina, which is prolonged over the villi, enwraps them all, and accompanies them even to the chorion.

To proceed in an orderly manner, I will now state what I have seen in the study of the villi. The observations of Farre are entirely exact, and confirm the doctrines of the ancients upon the cellular web, or external membrane, which surrounds the villi throughout their whole extent (Plate X., Fig. 2, *b, b*). By employing the customary processes of imbibition it is easy to perceive that, in placentæ at term at least, the cells are not inclosed in the substance of the membrane, but that they form, on the contrary, a species of epithelial layer which lines the whole internal surface of it. Exercising suitable pressure upon portions of the villi, it often occurs that the wall which surrounds the vessels, whence proceed the smaller villi, is torn. In this case, the epithelial layer, formed by the above-named oval and granular cells, is very easily perceived. I have never succeeded in meeting with such cells in the diaphanous substance that surrounds the vessels of the villi, in which, however, are clearly distinguished the fusiform cells which make a part of it (Plate X., Fig. 2, *a, a*).

I have stated above, that the exterior membrane surrounding the villi throughout their whole extent is furnished by the decidua serotina. I will now demonstrate it.

In animals as well as in woman, there takes place upon the internal surface of the uterus, where the ovum is developed, an abundant proliferation of connective tissue, the elements of which not only become more numerous, but each of them increases in volume. Between animals and the human species, however, there is this important difference: in the former, the hypertrophy and the hyperplasia of the sub-mucous connective tissue lift also the uterine mucous membrane, which is changed, as we have seen, into a glandular organ; while in woman, on the contrary, there is not only hypertrophy and hyperplasia of the anatomical elements constituting the connective tissue, but those elements undergo transformation and give rise to the development of a tissue of special neutral cells, which are the stroma whence arises the glandular organ that constitutes the maternal portion of the placenta. Instead of the epithelial layer which represents the mucous membrane, the uterus where the placenta is formed in woman remains covered by the serotina.

I owe to the kindness of our honored colleague, Dr. Belluzzi, the opportunity I have had to study the gravid uterus of a woman who died at the seventh month of pregnancy. Plate X., Fig. 3, represents a vertical section of the uterine surface of the placenta. At *a, a*, we see a layer of enormous fusiform cells, provided in part with short appendages. They show a hypertrophied and hyperplasic layer of cells of uterine connective tissue, of which we cannot say whether it belongs to the uterus, or is in process of transformation into the decidua serotina. It is certain that when in delivery the placenta is detached, its separation from the uterus is effected normally in this very layer.

On examining normal placentæ after their expulsion, and placing under the microscope the outer layers of the gray and jelly-like substance which covers their uterine surface, the very delicate cells which compose this layer are broken up, so that observing them thus, Robin was right in noticing their remarkable hypertrophy and their highly varied forms simulating those supposed to be characteristic of cancer, especially since he submitted to microscopic examination the pulp obtained by scraping. This process would necessarily detach thick round cells, with large nuclei, which form a second layer (Plate X., Fig. 3, *b, b*), namely, the serotina.

This part, whose cells are kept close together by interposed corpuscles of the connective tissue of the inferior layer I have described, presents the form of an elegant mosaic, on account of the rapid and remarkable metamorphoses which the cells of the serotina undergo. They are transformed into cells of fibrous tissue around the utero-placental vessels (Plate X., Fig. 3, *i*), and pass through the same transformation, in fulfilling their functions, as do real venous walls. About the venous sinuses which bear the blood from the placenta to the uterus (*h*), they are supported also against the villi of the chorion, and they have all around them a net-work of very delicate fibrous tissue which is the wall of the glandular organ; for, in its internal part, we see developed the lacteal cells of the uterus (*g*).

In favorable cuts from the base of the placenta we observe that, at the place where the villi are outwardly covered by the cells of the serotina changed into fibrous tissue, they are continuous in the interior (that is to say, in the serotina itself) with the lacteal cells of the internal surface. In a word, the internal surface of the cellular layer of the serotina is continuous with the internal layer of the wall which surrounds the villi.

At the base of the placenta the cells of the serotina are transformed into very apparent fibrous cells, and constitute quite a diaphanous fibrous tissue (c, c); in some places, as is represented in the illustration, may still be seen embedded the large round cells of the serotina. We also clearly perceive in the same figure, that the cells of the serotina (d, d) are changed into fibrous tissue and cover the villi with fibrous cells. If, higher up, in the external membrane of the villi, we no longer distinguish the fibrous elements, on account of its great transparency, nevertheless they are plainly seen at the place indicated, which is a continuation of the membrane which covers the villi in their smallest ramifications. The cellular or fundamental elements of the glandular organ thus line the whole internal surface of the fibrous envelope which the serotina furnishes to the villi (Plate X., Fig. 2, b, b).

The villi increasing in size and multiplying their subdivisions, and the volume of the placenta augmenting at the same time, the vascular loops which are formed push forward the primitive wall of the vascular branch or primordial villus, and we have the external wall which completely surrounds the villi. In this manner the glandular organ covers and envelops the subdivisions of the villi throughout their whole extent.

On approaching the chorion, the wall of the glandular organ which the serotina furnishes to the villi undergoes other modifications. Formerly granular, the internal cells become transparent; their form, which was oval, is rounded; and they fill the whole internal cavity of the glandular tube containing the vessels (Plate X., Fig. 1, d). Closely crowded together and in contact with the chorion, they are transformed into fibrous cells which, like large, tendinous cords, firmly fix the vessels to the chorion (e, e). This explains why anatomists thought that the chorion furnished a membrane to the vessels. Others seem to have been also in the right when they said that the serotina dipped down into the interior of the placenta, and that the cellular structure, indicated as its uterine surface (Plate X., Fig. 3, b, b), appears still unaltered upon the internal surface of the chorion which adheres to the fœtal surface of the placenta (Plate X., Fig. 1, f, f).

The modifications that I have always observed near the chorion in the internal cells of the glandular organ in its normal state (Plate X., Fig. 1, d), I have also seen in some pathological cases toward the base or uterine surface of the placenta. The large, quasi-epithelial cells which fill the internal cavity of the villi proliferate near the chorion in so remarkable a manner over the whole extent of the villi, that they prevent the blood from circulating to such a degree that some vessels are seen almost obliterated, while others have even disappeared.

By studying microscopically a placenta thus diseased, it is seen composed of tubes closely pressed together, whose walls are formed by the increased development of the glandular structure about the vascular villi, which are in great measure atrophied by the enormous proliferation of the cells which has been carried on in the internal wall of the glandular organ. The enlargement of the tubes is, in some places, so great that their exterior walls touch each other, and they leave in the utero-placental sinuses no space for the blood of the mother.

I speak of this single pathological lesion of the placenta, because it satisfactorily confirms the normal structure of the placental villi in the human species; and I point out thus the capital distinction between the human placenta and that of animals. Among the latter, we have seen that it is always a transformation of the uterine mucous membrane which gives rise to the development of the glandular organ or maternal portion of the placenta, without any departure from the ordinary typical forms of the most simple glandular organs. In the former, on the contrary, it is not a transformation or a perfecting of preëxisting parts, but the neoformation of a very special tissue, that of the serotina, which, if it has their fundamental characters, is yet widely removed from the typical forms of glandular organs in animals.

From these facts follows another which further constitutes a distinction between the human placenta and that of animals.

In the latter, the glandular organ originates from a transformation and expansion of the preëxisting mucous membrane, with hypertrophy and hyperplasia of the connective tissue and of the sub-mucous vessels; it follows that the villi of the placenta are always separated from each other by the walls of the follicle which receives it, and by the hypertrophied connective tissue interposed between the follicles, in the midst of which course the vessels that serve for nutrition. Consequently, in animals, the villi of the fœtal placenta are only and exclusively in direct contact with the fluid secreted by the glandular organ; in the human species, it has been in all times more or less well established that the villi of the fœtal placenta float in the maternal blood.

What I have just explained, upon the anatomical structure of the villi, must necessarily modify the general idea that has been formed with regard to them. In fact, the maternal blood directly bathes the external wall of the glandular organ furnished to the villi by the serotina. How is this accomplished? In Plate X., Fig. 1, g, g, g, I have represented the large cavities or lacunæ to be met with upon the internal portion of the uterine surface of the placenta, as well as throughout its whole thickness, even to the surface of the chorion, which, full of blood, as in the figure, and united with each other by intercommunications, constitute the cavities or placental sinuses, in which swim the villi entirely surrounded by the glandular organ.

Very different views are held by anatomists as to the manner in which the utero-placental vessels anastomose with each other to form these venous sinuses in the placenta.

I have specified the large and numerous cavities which are easily seen in the internal parts of the human placenta, in order to demonstrate that the external walls of the glandular organ, which surround all the villi, are continually bathed by the maternal blood. But this same observation leads me to believe that the utero-placental arteries empty directly into the lacunæ of the placenta as I shall presently explain.

I have not, however, yet been able to collect positive evidence sufficient to satisfy me in what portion of the placenta this takes place, and how the arteries act at the places where they open. My present purpose is to establish the fact that in the human placenta the vessels act quite differently from what we observe in animals. For with them, the external walls of the glandular organ are never bathed directly by the blood of the mother, and the utero-placental vessels are easily distinguished at all points of the connective tissue in the placental sections (Plate VIII., Figs. 1 and 2, *g, g*).

The most illustrious anatomists, failing to grasp the fundamental difference which exists between the human placenta and that of animals, and having often made use of that of the latter to explain that of the former, have necessàrily fallen into contradictions and errors which have been the cause of no slight confusion. This severe judgment upon such respectable observers will be pardoned me if I repeat the precise words of the celebrated Bischoff : —

" Hunter taught that in the same way as the decidua covers the remainder of the surface of the ovum, it covers it also as reflected decidua at the point where the placenta is formed ; but that in the course of time it attains a considerable development at that point, and it there forms numerous cavities with very thin walls, into which the villi of the fœtal part of the placenta insinuate themselves. He added that the uterine arteries and veins open into those small cells or cavities without branching or at most ramifying very little. The cavities are therefore always full of blood brought on one hand by the arteries, and taken away on the other by the veins. The later researches of Weber accord with those of Hunter upon essential points, with this one difference, that what the English anatomist calls cells of the decidua, Weber styles origin of the veins or venous sinuses. Consequently, while in the other parts of the body the arteries are divided into branches more and more slender, in order to be continuous, through the medium of the capillary net-work, with the equally minute roots of the veins; in the placenta, according to Weber, the uterine arteries are continuous, without furnishing any arborescent ramifications, with the sufficiently ample origins of the veins, which, anastomosing with each other frequently and at all points, appear to form thus a system of small cavities, to which the blood passes through venous trunks from the uterine arteries. The walls of the veins are extremely thin in the placenta; they are reduced to the single internal tunic, and when they do not contain blood, they contract, so as to become almost invisible. The villi of the chorion, constituting the fœtal placenta, which are divided into extremely minute ramifications, penetrate into the venous sinuses, where the delicate coating of the veins furnishes them with an envelope in the form of a sheath; so that they are always bathed by the maternal blood. Now, as the blood of the fœtus travels a long and quite sinuous passage through the villi, the two bloods find frequent opportunities for a reciprocal exchange of materials."

This description of Weber's, reproduced by Bischoff, has been pretty generally admitted in modern times. It agrees with the observations of Blokham, Knok,[1]

[1] London Gazette. 1840.

Reid, and Coste. But the researches of Eschricht raised doubts with regard to it. Basing his observations upon the placental structure of mammals, in which the two bloods of the mother and of the fœtus seem always to be conducted towards each other by capillary vessels, he concluded that in the human species, also, two net-works of capillary vessels continually come in contact, and that the uterine arteries are continuous with the uterine veins by means of a capillary net-work as tenuous as that which exists between the umbilical arteries and veins. He believed that prolongations of the decidua in the form of folds penetrate, in the interior of the placenta, between the ramifications of the chorion, and clothe them with a somewhat delicate membrane which serves as a support to the capillary net-work placed between the uterine arteries and veins.

This net-work has been imagined, it has not been seen, and the doctrine of Eschricht found no partisans; even a superficial inspection of the human placenta shows in its interior the large sinuses full of blood, in the midst of which float the villi. Hence Bischoff, as I have just said, attempted to reconcile the doctrines of Hunter and of Weber, and he was right. As for myself, it is sufficient, for the present, to have established that these illustrious men have been forced to admit, for the maternal vessels of the placenta, a different termination from the normal and ordinary one.

Notwithstanding my strong desire to form a clear and precise idea of the doctrine of Weber upon the circulation of the maternal blood in the placenta, I must frankly confess that I have never been able to succeed in comprehending how it could be that the villi of the fœtal placenta were bathed by the blood of the venous sinuses, since, in admitting, with the illustrious author, as thin a wall as one can imagine, in the dilated placental veins, one cannot be persuaded that the villi traverse it, and thus put themselves in contact with their blood.[1]

If the author meant that the very delicate venous walls are folded over the villi, enveloping them with a sort of sheath, the walls of the vessels of the fœtal villi would then necessarily be no longer in direct contact with the maternal blood, but rather with the wall of the vessel that contains them, and the cells and lacunæ which are quite distinctly seen could not be formed.

The researches that I have often repeated in order to follow the utero-placental vessels, venous as well as arterial, beyond the membrana serotina, have been unsuccessful, though the diameter of both is quite remarkable and renders them consequently very easily perceptible in that membrane (Plate X., Fig. 3, h, i). I am therefore disposed, as I have said before, to consider as correct the opinion of Farre, that the utero-placental arteries open directly into the sinuses of the placenta. But this subject deserves, in my opinion,

[1] I have the good fortune to possess the memoir of Weber, " Zusätze zur Lehre vom Baue und Verrichtungen der Geschlechtsorgane. Leipzig, 1846," which belonged to Eschricht, and I find the following words in it underlined with a pencil, counter-marked besides by an exclamation point: " In diese Mutterblute führenden Canäle insinuiren sich die zarten, gefässreichen, von Embryoblute durchstromtens Zotten des Kindestheils der Placenta, sie hängen daher in diese Canäle hinein und werden vom vorbeistromenden Mutterblute umspult." The doubts of Eschricht confirmed my own, without convincing me, however, of the existence of the very fine anastomoses he had imagined.

to be better studied and more fully explained. For if these arteries open upon the internal surface of the serotina, where the orifices of the large uteroplacental veins also exist, there would follow an outpouring of blood into the large cavities or lacunæ of the placenta, and there must be a continual mingling of the arterial blood of the mother with that which would have become venous.

I believe this is what really does take place; and I hope I shall not lack the opportunity of demonstrating it, when completing the observations which in the present work I have only alluded to. By this fact, the placental respiration would be explained in a simple and clear manner, and we could also account for another observation made by certain anatomists, that in the process of injecting the human gravid uterus, the coloring matters pass easily enough, if we operate from the fœtus to the mother; but this is not true, if we inject from the mother to the fœtus. In the first case, by lacerating any one of the villi, the coloring matter which pours out into the placental sinuses communicating with each other, and which from there insinuates itself with ease also into the large orifices of the arteries and of the utero-placental veins, thus passes without obstacle into the uterus.

I know no other example of lacunar circulation in any organ of the superior animals; and it appears to me that this of the human placenta signally deserves the attention of observers.

The existence of the utero-placental vessels, though formerly denied, is at present called in question by no one. Some doubt may still remain as to their genesis or mode of formation; and I do not pretend to resolve such a question peremptorily. I will only add that, in the study of the uterine mucous membrane of a dog almost at term, I was agreeably surprised to find a really splendid example for demonstrating the rich and abundant genesis of new vessels, by the vascular transformation of corpuscles of connective tissue. It is a phenomenon that one observes at the first glance, and which I have represented in Plate II., Fig. 1, *d, d*. The demonstration of the genesis of new vessels, by the method I have just indicated, not only serves to make these vessels understood, but is also usefully applied to several other questions of normal or pathological anatomy, and it furnishes a new and important argument to those who, like myself, believe that the corpuscles of the connective tissue are provided with a special wall.

At all events, the facility, on the one hand, with which we recognize, in vertical sections of the decidua serotina, the transverse cuts of the utero-placental arteries (Plate X., Fig. 3, *i*) and the large apertures of the veins (*h*), and their total absence in the portions of the serotina which become blended with the cells of which the chorion is composed (Plate X., Fig. 1, *f, f*); and, on the other hand, the easy demonstration of the large sinuses full of blood in the interior of the placenta, in which float the villi (Plate X., Fig. 1, *g, g*); the impossibility that has always existed of demonstrating the capillary net-work of the uteroplacental vessels in the placenta; the affirmations of Farre, and the numerous facts I have just stated, all lead to the conviction that the lacunar circulation, of which I have before spoken, is carried on in the placenta.

I do not pretend to have touched, even slightly, on all the very important questions which rise from the observations I have set forth. My aim has been to establish one single fact, namely, that the maternal portion of the placenta in the mammalia, and in the human species, has always a glandular structure. In the following chapter I shall sum up and arrange the statements deduced from this fact.

CHAPTER VII.

I.

CONCLUSIONS RELATIVE TO THE UTRICULAR GLANDS AND THE MUCOUS MEMBRANE OF THE UTERUS.

The uterine mucous membrane in woman and in certain animals, as the mare for example, is represented by simple epithelial layers. Small and narrow introflexions of the epithelial layer in some animals, in others, elevations of the sub-epithelial connective tissue, with numerous inflexions which form the folds of mucous membrane on the internal surface of the uterus, are not sufficient to establish real differences between the uterine mucous membrane of mammals and that of woman, still less to lead to the belief held by some illustrious anatomists, that the uterus of woman has no real mucous membrane.

The utricular glands of the uterus are usually very numerous, and open into the epithelial layer of the mucous membrane, whether that membrane shows itself totally inseparable from the uterine tissue, or whether it is lifted up as a special membrane and disposed in more or less prominent, simple, or highly festooned folds.

The great folds of the uterine mucous membrane, with their many festooned depressions, represent enormous glandular follicles which may stand for utricular glands. This fact is oftener remarked in certain animals in which the uterine glands are lacking. Some able anatomists have failed to observe them in the uterus of the rat, and I am convinced that they do not exist in that of the rabbit. The absence of the utricular glands in the uterus of certain animals with a single placenta is a fact of no slight importance, for it invalidates in some degree the assertion of those who claim that these glands play an important part in the formation of the placenta in the above-named animals.

In those animals in which the uterine glands have been studied with attention, quite remarkable differences are noted, in relation both to their form and to the kind of epithelium which lines their cavity. Since the observations of Sharpey and Weber, it has been admitted as a demonstrated fact in science that in the uterus of some animals, such as the cat and the dog, there exist two species of uterine glands, which, on account of their form and volume, have been named simple and branching.[1] Moreover, small and not important differences are noticed in the utricular glands of the uterus in all animals and even in the human species. Hence the error of the anatomists and physiologists who claim that there are two species of glands with a double and widely differing function, namely: the secretion of the uterine mucus for the simple glands, and a participation in the formation of the placenta for the branching ones.

I have, however, observed two really distinct species of uterine glands in the cow and the sheep: the

---

[1] I have been unable to demonstrate that there are two varieties of these glands in the dog, and I have shown that in the cat there is only one species of glands, although they may vary greatly in volume.

utricular or branching, of slightly variable volume, but always highly developed, and the simple glands, always extremely small, proceeding from very narrow and sinuous inflexions of the epithelial surface of the mucous membrane. Moreover, these minute, slightly developed glands, which, to distinguish them from the others, I have called glandular follicles, present, on comparison with each other, marked differences as to length and breadth. They are scattered over the whole internal surface of the uterus, and are constantly agglomerated in the places corresponding to the cotyledons, which, in the non-gravid uterus, appear covered with a thin, smooth, and compact layer of epithelium; this represents the most simple form of the uterine mucous membrane, as in woman.

In the rabbit, instead of utricular glands, we find over the whole surface of the uterine mucous membrane numerous and very short glandular follicles, which are only inflexions of the epithelial layer, representing also, in that animal, the mucous membrane of the uterus. In this case the only difference is that the internal surface of the mucous membrane does not appear glossy and smooth as in woman.

In all animals in which the uterine utricular glands exist, as well as in the human species, they increase in volume during pregnancy. The glandular follicles also augment in size during the gestation of the cow. The development of the glandular follicles in the gravid uterus of the rabbit is still more remarkable, and it has a much greater importance and significance. At the places where the ova are arrested after their fecundation, the follicles augment in volume and are transformed into a glandular organ or the maternal portion of the placenta. In the empty portions of the uterine horns the growth of the follicles causes the elevation of the mucous membrane under the form of peculiar folds; they seem destined to perform, during pregnancy, the functions of the utricular glands, which are lacking in those animals.

In cases where the placenta is villous or diffused, as in the mare, all the utricular glands, even after the glandular organ or uterine portion of the placenta is formed, empty the fluid which they elaborate directly into the space comprised between the chorion and the uterus. The uterine surface of the chorion of these animals is covered with an epithelial layer which clothes also the base of the tufts of its villi, and is continuous with the epithelium that covers them. The outer epithelial layer of the chorion may represent the uterine decidua in the mare.

When the placenta is multiple, as in the ruminants, and especially in the cow, the uterine utricular glands, which do not correspond to the cotyledons, likewise pour their fluid between the chorion and the uterus. The epithelial layer which forms the decidua of the cow is a little more remarkable than in the mare. The utricular glands which exist in the so-called rudimentary cotyledons of the non-gravid uterus, as well as the glandular follicles agglomerated in those parts of the organ, probably open into the bottom of the cup-shaped elevations which constitute the new-formed glandular portion of the cotyledons of the gravid uterus. The small number of utricular glands at that point, the delicacy of the mucous follicles, and, still more, the increased attenuation of the walls of the glands, the transparency and the metamorphosis of their internal epithelium, never allowed me to see the precise point where they open in the interior of the glandular organ. The glands and the follicles, which are clearly seen in transverse sections of the peduncle of the cotyledon, are badly and imperfectly distinguished in vertical sections, and they become less and less marked the nearer we come to the surface of the peduncle where the glandular organ is formed.

If the placenta is single, and if there exist utricular glands, as in the carnivora, those which correspond to the place where the placenta is developed open into the lower part or *culs-de-sac* of the glandular follicles of new formation, which are nothing but festooned folds of the uterine mucous membrane transformed into a glandular organ. In the rest of the uterus, even in these animals, the utricular glands pour the secreted fluid between the uterus and the chorion.

The uterine decidua of woman, as well as the so-called catamenial deciduæ, is a product of materials elaborated by the utricular glands. Consequently, the decidua cannot be considered as a swelling of the uterine mucous membrane, and still less as resulting from the extremities of the glands, from the connective tissue, and from the vessels that surround them, as Weber and Bischoff have stated. The numerous openings or holes, which give the appearance of a sieve to the uterine decidua of the human species, only indicate the points corresponding to the orifices of the utricular glands in the cavity of the uterus which remain fully open to allow the constant passage of the product of their secretion.

The uterine decidua may be demonstrated also in the cow, though several authors have denied its existence, on account of its attenuation. It has the same origin as the human decidua, but because in the cow, besides being thin, it is attached to the chorion and not to the uterus, as in woman, we do not observe in it the numerous openings that are seen in the latter. In the vaccine decidua, instead of holes, there is a thickening of some of the elements which are secreted by the glands at the points corresponding to their orifices, which infiltrates as far as the chorion. Burkardt gave them the name of chorial *squamulæ*. These opposite conditions confirm the fundamental fact concerning the origin and structure of the uterine decidua.

In no species of animals, whatever the form of the placenta, do the villi of the chorion penetrate into the utricular glands of the uterus, as some anatomists have taught.

The constant increase in volume of the utricular glands during pregnancy, in animals as well as in the human species, proves beyond doubt that they have an important function to fulfill for the life of the fœtus. For the present, it appears to me reasonable to suppose that their principal function is that of furnishing materials for its nutrition, before the development of the new glandular organ which constitutes the maternal portion of the placenta in all mammals and in the human species. Although the fluid secreted by the utricular glands does not always mingle directly with that which is elaborated by the maternal placenta, as in the carnivora, the undeniable observation of this fact in some animals leads us reasonably to suppose that an important nutritive element is furnished by these glands for the nutrition and growth of the fœtus. This seems more probable still, if we think of the very great number of these glands, of their constant increase in volume during pregnancy, and of the remarkable quantity of fluid they secrete in some animals, as in the mare, between the chorion and the uterus; and finally, if we take into account the fact that the whole uterine mucous membrane augments in volume during gestation, and multiplies its excavations or festoons, which exhibit enormous follicles in those animals in which the true utricular glands are wanting.

II.

CONCLUSIONS UPON THE GLANDULAR ORGAN OF NEO-FORMATION OR MATERNAL PORTION OF THE PLACENTA, IN THE MAMMALIA AND IN THE HUMAN SPECIES.

In the uterus of all mammals, woman included, there is formed during pregnancy a new glandular organ, into the internal cavities of which the villi of the chorion always penetrate.

The placenta is, therefore, formed of two parts, entirely distinct both in structure and in function: the fœtal portion, which is vascular or absorbent, and the maternal portion, which is glandular or secretory.

The blood of the mother always furnishes the elements for the formation and secretion of the new glandular organ or maternal placenta. In no case do the maternal vessels cross each other and come in contact with those of the fœtus; or, in other words, the parts constituting the fœtal placenta are always in contact with the fluid elaborated by the new glandular organ, and are bathed by it.

The doctrine, universally admitted by physiologists, of the fœtal nutrition by means of an exchange of materials through the processes of endosmose and exosmose between the vessels of the mother and those of the fœtus fails in the presence of actual observation. In the same manner as in the first periods of extra-uterine life the child is nourished by the maternal milk absorbed by the intestinal villi, so during intra-uterine life the fœtus finds its nourishment in the fluid or milk secreted by the glandular organ and absorbed by the villi of the chorion. Anatomical researches have led to the discovery of the simple fact. Physiology and chemistry will make known the secrets of function.

The new glandular organ or maternal portion of the placenta is developed at different periods of pregnancy in different species of animals. In case the placenta is diffused, as in the solipeds, it appears over the whole internal surface of the uterus. When the placenta is multiple, it develops at certain circumscribed points, as in the ruminants. Finally, it forms at the place where the ovum is arrested, when the placenta is single, as in the rodents, the carnivora, and the human species. The growth of the new glandular organ is modified, according to the various forms of the placenta; but in animals it does not differ from the simplest types of glandular organs in adult individuals. In short, in animals the maternal placenta always preserves the character of an open glandular follicle. The typical form of the simplest glandular follicle is, however, wanting in the human species.

The anatomical cause of the differences between animals and the human species lies in the fact that, in animals, the new glandular organ, or maternal placenta, results from a modification and transformation of the preëxisting mucous membrane of the uterus; while in woman the same portion of the placenta is formed by a stroma, which stroma is itself of new development, and is elaborated from the connective tissue of the internal surface of the uterus. This stroma is known by anatomists under the name of decidua serotina.

The fundamental type of this new glandular organ is that of single follicles crowded close together and lining the whole internal uterine surface, as we see in animals with a disseminated placenta, and as I have demonstrated in the uterus of the mare at term. In animals with a multiple placenta, such as the ruminants, and particularly in the cow, as I have shown, a clear distinction has not hitherto been made between the permanent part of the uterine cotyledons and the glandular portion of new formation (that is to say, the caducous or perishable part, which disappears after delivery), which is developed upon the cotyledons only during gestation. The persistent portions, which are observed even in the fœtus, and which are called rudimentary cotyledons, simply indicate the places where, during pregnancy, the maternal or glandular portions of the placenta will be developed.

In the cow, the new glandular portion of the cotyledon preserves the form of an aggregation of simple, open follicles. But if we compare them with the cotyledons of the mare, we find the only difference to be their manner of union and position in the uterus. In the cow they are not simply glandular follicles lying near together, but they are superimposed upon each other, and they are not found over the whole uterine surface, but only upon the portions of the non-gravid uterus called uterine cotyledons. They are not vertical, like the former, but parallel to the line formed by the uterine surface. Neither do they open separately, nor directly into the cavity of the uterus, but indirectly by means of one large aperture corresponding to an internal cavity, into which many follicles empty themselves in common.

It still remains to be known how and by what histogenetic process the new and glandular part of the cotyledons is developed, and at what period of pregnancy the glandular follicles are formed in the gravid uterus of the solipeds. Observations made upon the carnivora, which I shall presently describe, lead us to take these processes for granted, as they also allow us to infer the manner in which the new glandular portion is destroyed after delivery, in the mare as well as in the cow. But an exposition of the facts, when ascertained, will be better than conjecture, however probable.

In animals with a single placenta, such as the rodents and the carnivora, the glandular organ is strikingly modified by the forms which it assumes, without losing those characteristics which belong to the fundamental and simple type of a follicle. The changes which it undergoes relate only to the length and to the extremely sinuous course of the glandular follicles, and to the multiple communications which they have with each other. But the closed extremity of the separate follicles of the uterine face of the placenta, and their orifices towards its fœtal surface, are always easily observed. In all cases the chorial villi of the fœtal placenta penetrate through orifices into the interior of the follicles; only, when the placenta is single, the chorion adheres to its fœtal surface.

The modifications I have pointed out in the new glandular organ of the uterus of mammals, according to the various forms of the placenta, recall and demonstrate, in a higher degree of development, the differences described in the mucous membrane of the non-gravid uterus. The slight and narrow depressions that are met with in the mucous epithelium of some animals are reproduced on a large scale by the structure and disposition of the maternal placenta in the solipeds. In like manner, the fold-like elevations of the mucous membrane, with numerous and broad lateral sinuses, which we have remarked in other animals, are represented, under a more complex form, by the glandular portion of the cotyledons in the gravid cow. On the other hand, the long and sinuous follicles of the maternal or glandular portion of the single placenta in certain animals show only a remarkable augmentation of the follicles and depressions preëxisting in the uterine mucous membrane of other animals.

The maternal portion of the placenta remains intact in the uterus during delivery, and is then destroyed gradually, in cases where the placenta is disseminated or multiple. In the mare there remains no trace of it on the internal surface of the non-gravid uterus. In the cow it is preëxistent at pregnancy; after delivery, we find the traces of the points where the new glandular organ was formed, and where it will be formed again in successive pregnancies. The trace of these places is known under the name of rudimentary cotyledons, even in the uterus of the fœtus.

In case the placenta is single, the portion of the uterus which was occupied by the placenta remains, after delivery, covered by a thickened mucous membrane, surrounded on the sides by very elevated folds.

This thickened mucous membrane, quite wide and somewhat rough a few hours after delivery, appears three days later lifted into several closely-crowded folds. These changes should be attributed to the return of the uterus to its normal state. In delivery, only that portion of the elevations of the mucous membrane is detached which had been transformed in order to constitute the new glandular organ or maternal placenta. The portion of the mucous membrane remaining in the uterus, and whose changes I have described, is itself gradually and entirely destroyed by fatty degeneration. Thirty days after parturition I found it had completely disappeared from the uterus of a dog.

In the human species only is there a total separation and expulsion of the glandular organ in delivery. Consequently, in woman alone does there take place an extensive traumatic lesion of the uterus, on account of the laceration of the parts, which leaves bare the uterine tissue over all that portion which had been covered by the placenta. In animals with a single placenta this lesion is limited to the connective tissue of the folds of the mucous membrane which has accompanied the elevation and the growth of the follicles of new formation. The uterine contractions, the drawing together of the parts by the diminution in volume of the uterus, bring prompt and efficacious relief to this lesion. There is an additional anatomical explanation, from these facts, of the sentence addressed to woman: "In sorrow thou shalt bring forth children."

Important differences are observed in the maternal placenta of the human species which remove it from the type common to animals. In woman it is not the uterine mucous membrane that is perfected, as in animals, in order to form the glandular organ, but it arises from the neoformation of a layer made up of large cells furnished by the sub-mucous connective tissue of the uterus, and known by anatomists under the name of decidua serotina. The large, neutral cells of the serotina are the stroma whence the maternal or glandular portion of the placenta takes its origin.

The fundamental and typical parts of the glandular tissues are met with in the maternal portion of the placenta in woman. All the accessory characteristics, that is to say, those which relate to the form of a simple glandular follicle, disappear altogether.

The cellular structure of the serotina which lines the uterine face of the placenta is also observed with ease upon its fœtal surface covered by the chorion, where the cells of the serotina are blended with the connective tissue. It is, therefore, demonstrated in a clear way, that the serotina penetrates into the interior of the placenta. There the cells of the serotina are transformed at different points into true fibrous tissue, more especially to circumscribe the large lacunæ of the placenta which contain the maternal blood. The same transformation takes place in the substance of the serotina, to furnish a solid wall to the uteroplacental veins before they reach the uterus. Further, the serotina clothes the villi of the chorion throughout their whole extent and in their numerous ramifications to the interior of the placenta. Everywhere the cells of the serotina offer examples of the greatest and most rapid modifications. The most important consists in the sheath which the serotina furnishes to the villi of the fœtal placenta as far as the chorion. This sheath is formed on the exterior by a fibrous membrane, and by an internal epithelial layer, which together constitute the fundamental parts of the glandular organs.

Near the chorion, the decidua serotina is transformed by degrees into fibrous tissue, and forms strong cords which serve to fix firmly the vascular trunks, whence depart the chorial villi. I have observed that this same thing had taken place, in an abnormal manner, in a morbid placenta, which produced the death of the fœtus and caused abortion.

When once the vascular trunks of the fœtal placenta are enveloped by the serotina, transformed into a secreting glandular organ, the numerous villi that depart from it push before them, while increasing in volume, the walls of the sheath, and they are themselves thus completely clothed with it, like the fingers of a hand in a glove. In this way the blood of the mother bathes directly the exterior wall of the sheath furnished by the serotina to the villi.

In the human species only the utero-placental arteries and veins are not divided into trunks and branches in the placenta. The maternal blood is poured out, in the interior of the placenta, into large cavities, lacunæ, or sinuses, which communicate with each other, and are circumscribed by the chorion on the fœtal side and by the serotina on the uterine side. The cavities circumscribed by the decidua serotina, transformed, at those points, into fibrous tissue, are in great measure filled by the blood and by the voluminous tufts of the chorial villi that are covered by the serotina, changed into a glandular organ.

The intimate union of the vessels with the chorion and the serotina, the internal prolongations of the latter, which are mingled with those of the chorion, limit the distention of the internal cavities or placental lacunæ, — a distention which would necessarily be produced by the arterial blood which constantly comes from the mother to the placenta. This blood, which is poured into these lacunæ, is mingled with that which has already become venous in the interior of the organ. The great cavities or venous sinuses, circumscribed by the serotina, carry back to the mother the blood which has fulfilled its office in the placenta by means of the utero-placental veins.

In the human placenta, also, the vessels which bring the maternal blood never come in contact with those which belong to the fœtus.

So it is only in the human species that a mixed blood, arterial and venous, is brought in contact with the external face of the glandular organ which contains the vessels of the fœtus; and this by a mode of lacunose circulation, of which, to the present day, no example has been met with in the superior animals.

It is in the human species alone that a mixed blood of the mother herself is brought back from the placenta into the general circulation.

APPENDIX.[1]

I.

ON THE FORMATION OF THE MATERNAL OR GLANDULAR PORTION OF THE PLACENTA IN THE HUMAN SPECIES AND IN CERTAIN ANIMALS.

WHEN Professor Bruch and my old friend Andreini proposed to me to publish my memoir upon the placenta in French, I had at first the idea of revising it, at least in part; for new researches and later observations had shown me that it needed to be somewhat modified, or to have some additions made to it. I afterwards decided to have it followed by an appendix, in which are to be found the new observations and the modifications resulting from them.

Young persons who devote themselves to the study of the natural sciences may thus be convinced that the investigation of facts is not an easy thing, as many seem to think, but that it is, on the contrary,

[1] This Appendix, written for the French edition, was thoroughly revised and published in the Transactions of the Academy of Sciences of the Bologna Institute, second series, vol. ix., and illustrated by six plates. At first I thought to present it in place of the Appendix to the English reader. However, decision was made to retain the Appendix, and publish entire the professor's more recent monograph on the Unity of the Anatomical Type of the Placenta in Mammals and in the Human Species, and on the Physiological Unity of the Nutrition of the Fœtus in all the Vertebrates, published in the Transactions of the Institute, 1877. — H. O. M.

always long and often laborious. For that very reason it is generally fruitful in valuable results.

While truthfully presenting observations, one may sometimes be mistaken in the manner of interpreting them; however, the fact remains entire in its truth, and sooner or later it becomes useful to science.

For example, in having drawings made of the large excavations in Plate IX., *d, d*, I regarded them as *culs-de-sac* of large folds of the uterine mucous membrane, upon which the maternal placenta of the dog had been formed. It was an error. Later observations have proved to me incontestably that these large cavities are produced by a degeneration and special dilatation of the subjacent utricular glands at the place where the placenta is developed. The fact remains; its interpretation only is modified. In order to appreciate it fully, one must have followed it from its discovery to its development.

It is thus that we teach ourselves and others, and throw light upon science, which is pure and simple truth.

I hope, then, that not only for young students, but for all others likewise, it will be of great service to follow conscientious observers in the path which conducts them from error to truth, and I trust at least that my purpose will not be criticised.

In my memoir I have scarcely touched upon the investigation into and the description of the histogenetic process and the successive transformations of the primordial elements constituting the glandular organ, or maternal portion of the placenta.

Such a research is pursued with disadvantage, on

account of the difficulty one is under of procuring gravid animals at different periods of gestation. It is, however, of the greatest interest; for it is only by the exact and precise knowledge of those successive changes that a clear and certain proof of the neoformation is deduced, and that the opinion can be refuted of those who, not having had the opportunity of repeating minute and long-continued observations, think that the glandular organ is only an expansion, or a simple transformation, of preëxisting uterine parts, as has been said of the utricular glands.

I shall therefore summarily describe in this Appendix the results of my studies made after the publication of the memoir. I shall limit myself to those which suffice to prove that, during pregnancy, there takes places a true glandular neoformation, and that, moreover, the glandular organ in its mode of development wholly departs from the ordinary and known laws which regulate the formation of glands in the animal organism.

The permanence of the glands in beings that are organized and destined to live, and the necessarily temporary existence of the glandular organ of the placenta, reveal the purpose, if not the explanation, for the diversity of the laws which, in each case, direct their formation.

Histologists agree in saying that all glands develop by means of an introflexion of an epithelial layer, and that the differences to be observed among them are only more or less remarkable modifications of this primitive and constant fact. In the formation of the maternal placenta, the glandular organ, though it begins in different ways in the different species of animals, is never formed by an introflexion of the uterine epithelium and of the sub-epithelial connective tissue. It is constantly the result of a production of histological elements unlike those that existed, and their successive changes constitute the glandular portion of the placenta. The placenta is destined to be loosened and expelled from the uterus in parturition, or slowly to degenerate and afterwards disappear, according as the maternal portion remains entirely attached to the uterus, as in the cow, or only partially so, as in the dog.

I shall therefore describe these modes of placental formation in the cow, among the ruminants; in the cat, among the carnivora; and in woman. I shall add to this the observations upon certain peculiarities of placental structure in animals that I had not before been able to examine.

II.

ON THE FORMATION OF THE MATERNAL PLACENTA IN THE COW, AND OF THE PLACENTA OF THE SHEEP, THE MOLE, AND THE HIND.

SECTION I.

THE COW.

AMONG the ruminants, the animal which has offered me the greatest facilities in my researches is the cow. The public abattoir has furnished me all the uteri of those animals killed in a condition of pregnancy; and, notwithstanding the agricultural regulations of our country, which are opposed to their being thus slaughtered, I have been able to obtain two, of which the fœtuses were found at that degree of development usually assigned to between the fiftieth and fifty-fourth days of gestation.

At this period of pregnancy, the glandular organ, though small, is already completely formed in the middle part of the full uterine horn. Below, and still better towards the extremity of the horn, smaller cotyledons are seen at different degrees of development, which are very well adapted to the study of the various phases of the neoformation of the glandular part of the placenta, that is to say, of the uterine cotyledon.

To be brief, I shall review the facts that I have observed; and, to make them as clear as possible, I shall present, at the different periods of evolution, the most remarkable changes that I have noted, from the most simple or primordial to the formation of the large calices represented in Plate VI.

FIRST PERIOD. — We perceive only a slight tumefaction over the whole surface of the uterine cotyledons, — a surface which was level and smooth in the nongravid uterus. According to the position of the uterus, the swollen surface of the cotyledons has the appearance of a soft crust, of a yellowish-white in some, and of a bright red in others. By the simple pressure of the uterine walls which touch the cotyledons, it is easy to make those which were yellowish turn red, and *vice versa.* This makes it certain that the different coloration of the cotyledons does not depend merely upon the fullness or emptiness of the blood-vessels; it proves also that the facility with which the two conditions are produced is connected with the extent of the vascular net-work which has been developed in the sub-epithelial connective tissue of the uterine surface of the cotyledons.

If, for this experiment, we make use of a magnifying-glass, we easily distinguish this vascularization. We also see the large sinuosities of the more superficial new vessels, and we perceive that their convexities are turned towards the uterine cavity.

Consequently, if we examine with the microscope a vertical section of these cotyledons, it no longer appears even and smooth, as in the unimpregnated state, but everywhere slightly undulating. The first period of the neoformation would then be that of a vascular hyperplasia.

This vascularization may be rapidly established, and in Plate II., Fig. 1, *d, d,* I have already shown how it is produced in the uterus of the dog.

The easy demonstration of the direct communication of the corpuscles of connective tissue with the blood-vessels does not merely serve to explain the rapidity of the vascular neoformation, but it is also, I think, a beautiful method for illustrating the passage of the white globules of the blood into the corpuscles of the connective tissue.

Cohnheim has, by his observations, forcibly called the attention of pathologists to this fact.

SECOND PERIOD. — At this stage of the neoformation of the glandular organ, one may easily understand the sinuosity and the curvature assumed by the vessels of the sub-epithelial net-work of the cotyledons towards the internal cavity of the uterus.

THIRD PERIOD. — Between this and the second period there is no essential distinction, but simply a difference of volume or development in the vascular loops. They lose, in fact, the tenuity which they had on the old surface of the cotyledon, in order to form projections, rising and surrounding themselves with connective tissue, and becoming covered with a thin epithelium. They are thus organized into small and delicate villi, crowded close together.

These simple papillæ will become, in the fully formed glandular organ (Plate VI., *a*), the columns of connective tissue which rise and constitute the walls of the calices.

FOURTH PERIOD. — We may assign to this stage the proliferation of these simple villi into other lateral villi.

These lateral villi will be represented in the completely formed cotyledon (Plate VI., *b, b,*) by the numerous horizontal follicles of the interior of the calices.

FIFTH PERIOD. — This comprises the union of the lateral ramifications of a villus with those of the neighboring villi.

SIXTH AND LAST PERIOD. — It is now that, after the union of the above-named ramifications of the villi, the calices of the new glandular organ are complete, as well as the internal follicles that are superimposed upon each other. They are illustrated by diagram in Plate I., Fig. 2, and represented according to nature in Plate VI.

SECTION II.

THE SHEEP.

In the memoir I scarcely touched upon the anatomical structure of the cotyledons in the sheep. I limited myself to the assertion that the differences between the cotyledons of cows and those of sheep were more remarkable and more important than had generally been stated, specifying the concave form in the one and the convex in the other.

Now I am able to add that the cotyledons of the gravid uterus of sheep are formed on the interior by trabeculæ clothed with epithelium. They rise irregularly from its internal uterine walls, and are distributed without order into lateral ramifications, which adhere to and cross each other, thus leaving spaces or intermediate and irregular cavities which are occupied by voluminous tufts of chorial villi.

It is thus that, while it exists and performs its functions, the glandular organ of the sheep preserves the forms indicated for the fifth genetic period of the cotyledon of the cow.

Briefly, the glandular organ of the sheep represents a normal arrest of the successive development of the glandular organ of the cow; it allows us to see constantly and with the unaided eye what in the latter animal may be observed by the aid of the microscope for a limited period.

SECTION III.

THE MOLE.

The simplest form of the uterine cotyledon that I have pointed out in the sheep is also observed in an animal with a single placenta. This animal is the European mole. Such a fact seems to me worthy of special mention. It not only brings the form of single placenta near to the cotyledons of the ruminants; it does still more.

In an interesting work of Professor Bruch, of Strasburg,[1] with which I was not acquainted at the time my memoir was published, it is very clearly demonstrated and represented by plates that among the inferior vertebrates, such as the squalidæ, the internal uterine surface is clothed during pregnancy with an infinite number of vascular villi covered with epithelium.

In these animals, consequently, the maternal portion of the placenta would be represented by the first period of formation that I have noted in the cotyledons of the cow.

SECTION IV.

THE HIND.

The anatomical differences that I have indicated are not the only ones to be met with in the gravid uterus of the ruminants; and I do not intend even to name them all.

I have had, however, the opportunity to study a cotyledon of a hind (*Cervus axis*), and I have assured myself that the differences between it and those of the cow are not limited, as Harvey says, to the less number in the hind, nor to their reputed smaller size.[2]

[1] Études sur l'Appareil de la Génération chez les Sélaciens. Strasburg, 1860.
[2] See Memoir, page 51.

The cotyledon of the hind that I have been able to observe has the form of a kidney. Its longitudinal is much greater than its transverse diameter. The first measures a little less than eight centimeters; the second, a little more than five; the circumference twenty-one, and the thickness three and one half centimeters.

This voluminous cotyledon arises from a narrow fold of the mucous membrane in the connective tissue, of which we perceive a rich net-work of large vessels. From the surface of this fold there is developed a great number of tubes or glandular follicles crowded close together, of a length corresponding to the thickness of the cotyledon.

These follicles are tunnel-shaped. The transverse sections show the different diameters, ten or twelve hundredths of a millimeter at the orifice which gives entrance to the villi of the chorion, and two or two and a half at their base. These transverse sections present the exact figure of a honeycomb with round cells.

One can form a precise idea of the structure, as well as of the anatomical elements constituting the glandular portion, of the placenta of this species of deer by looking at Plate V., Fig. 1, *b, b*, which represents a transverse cut of the maternal organ in the mare.

III.

ON THE FORMATION OF THE SINGLE PLACENTA IN THE CAT, THE HARE, AND THE GUINEA-PIG.

In treating of the anatomical structure of the single placenta in the mole, I have already pointed out the analogies which it has with the cotyledons of a ruminant, — of the sheep, for example.

I have since met with other differences in various single placentæ of certain animals. These conditions, which I could not elaborate in my memoir, further confirm, by their great variety, the single type of the glandular organ.

Here I shall content myself with reviewing the mode of formation of the placenta of the cat, and I shall add to this review some observations upon that of the hare and of the guinea-pig.

They will serve to demonstrate that if a veritable and actual neoformation of the glandular organ is constant it may yet take place by means of a histogenetic process different from that which I have described in the cow, namely, by the production of a special cellular tissue formed of cells proper, analogous to those I have represented in treating of the decidua serotina of the human species, in Plate X., Fig. 3, b, b, and Fig. 2, f, f.

In the edentata, also, the glandular organ is the product of the successive transformation of these cells; and in that case, as well as in man, this phenomenon takes place contrary to the ordinary laws which govern the formation and the development of the permanent glands in living beings.

The existence of the decidua serotina in animals had been denied, as I have already remarked. I have pointed out the importance of it in the human species.

In following the mode of formation of the placenta in the cat, I have become convinced that, in that animal also, there takes place upon the surface of the folds of the uterine mucous membrane, at the place where the placenta is formed, a neoformation of very delicate cells, which finally, after successive transformations, constitutes the new glandular organ.

I have stated that the same thing takes place in woman, though in its forms the glandular organ presents the same differences that I have noted between the placenta of woman and that of the dog.

What I now add is, that in these two carnivora, the dog and the cat, the neoformation has the same origin as in woman, that is to say, in the special cellular elements which represent the modified serotina.

A very important peculiarity will be offered us in connection with this subject by the placenta of the hare, of which I shall treat after having briefly indicated the formative process of the placenta of the cat.

SECTION 1.

THE CAT.

The uterine mucous membrane of the cat undergoes, during pregnancy, remarkable modifications, both at the places where the ova are arrested and in the portions of the uterine horns which remain empty.

I shall notice these changes separately, beginning

with those which are observed in the empty portions of the uterine horns.

It was already known that the mucous membrane of the uterus is swollen during the rutting period and the period of pregnancy. The changes which it presents during this tumefaction have not been specified, that I am aware of, with the exception of a greater flow of blood and the expansion of the utricular glands, remarked by Malpighi.

Now my observations upon the cat have demonstrated to me that after the tenth day of pregnancy the uterine mucous membrane is so tumefied in the vacant portions of the uterine horns that the result is a complete obstruction of the cavity of the uterus. In this way the ovum remains confined at the point where it is arrested, in a cavity closed in upon all sides. In proportion as the ovum is developed, and as the placenta is organized, forming relations of a certain duration between the ovum and the uterus, the cavity of the horns is reëstablished, and that not in consequence of the decreased hypertrophy of the mucous membrane, but because the increase in volume of the uterus enlarges the whole circumference of it, and the cavity of the horns also shares in the enlargement.

The epithelial layer of the uterine mucous membrane, which forms a smooth and lubricated pellicle (*velamento*) in the non-gravid cat, is transformed, by means of the vascular hyperplasia and by the sub-epithelial connective tissue, into folds which present, during gestation, the appearance of voluminous glandular follicles, like those of the uterus of the rabbit, as represented in Plate VII., Fig. 1. Upon the increase in volume of these folds depends the complete occlusion of the cavity of the uterus at the places where the ovum is not arrested.

With the progress of the pregnancy and the reëstablishment of the cavity, it is easy to distinguish that the folds present at the sides numerous festoons, a disposition that Leydig affirmed to be normal in the non-gravid uterus of certain animals.

In recalling these observations in my memoir, I also remarked that that anatomist had suspected that these great festooned folds of the uterus of certain animals might represent under an enormous development the extremely small glandular follicles or mucous crypts to be met with in the uterine mucous membrane of other animals.

The changes that I have named in the uterine mucous membrane of the cat represent successively the most remarkable forms and differences that anatomists have found to exist in the uterine mucous membrane of mammals, and it appears to me that observations confirm the supposition that Leydig brought forward when writing philosophically upon the subject of anatomy. At all events, these observations lend probability to it, and I believe support me also, when, according to Bischoff, I considered the epithelial layer which lines the uterus of woman, and the cotyledons of the non-gravid uterus of the cow, as the simple and elementary form of a mucous membrane.[1]

At the place where the ovum has been arrested the uterine mucous membrane of the cat assumes at first a follicle-like appearance, as is represented in the

[1] See Memoir, page 41.

rabbit (Plate VII., Fig. 1). The folds, and consequently the depressions, are very small. Where the placenta is not formed they promptly disappear, and the mucous membrane becomes again smooth on account of the distention which the increase in size of the ovum produces upon the uterine walls. At the point where the placenta is formed, on the contrary, the epithelium which covers the slight follicles that have been developed seems to become soft and to take on a tomentose appearance. At the same time, from the sub-epithelial connective tissue, proliferates another tissue of rounded, soft, and delicate cells which blend with those of the softened epithelium. The form of the follicles is preserved by the elevation of this tissue of neoformation into straight, slender lamellæ, at first vertical and covered with a delicate epithelium which corresponds to that which lined the uterine mucous membrane.

Between these laminæ produced by the cells of new formation are insinuated laminated prolongations of the chorion, where, only later, do we distinguish the vessels.

During the progress of the formation and the development of the maternal portion of the placenta the lamellæ are elongated without increasing in thickness, and, under the pressure of the growth of the ovum, they fold and refold over each other, until they present precisely the structure of the perfected glandular organ, as I have demonstrated in the placenta of the dog (Plate VIII., Fig. 2).

There is, however, a capital difference. During this period of formation the laminæ of the new tissue are single, but when the glandular tubes shall be formed, as in the above-named figure, each tube will be composed of the half of two folds which are united together, inclosing the vessels that have been formed in the chorial laminæ, interposed from the very origin of the placenta between the lamellæ of neoformation.

The volume of the folds which rise from the uterine connective tissue does not differ from that of the completely formed glandular tubes. It follows that it is easy, at first view, to take one for the other, and to confound a fold, sinuous in shape, with a complete follicle having the same kind of conformation.

But it is not difficult to avoid this error by observing with attention the distribution of the utero-placental vessels, whose canalization is soon distinguishable in the midst of the cells of the new-formed laminæ; on the contrary, the vascularity of the chorial laminæ, that is to say, the fœtal portion of the placenta, is not yet formed.

Continuing the observations upon the progressive development, it is also easily seen that while the vascularity of the laminæ of the chorion is being established the vessels are surrounded by the laminæ of new formation, and the half of a fold blending with that half which is in contact with it constitutes the long and sinuous glandular follicle which I have described in the placenta of the dog, and which is not different from that of the cat.

The mode of union of the uterine laminæ, which form the glandular follicles and surround the vessels of the chorion, explains the numerous communications which the vessels of the fœtal placenta preserve with each other (Plate VIII., Figs. 1 and 2, *e, e*), which remain, however, all included in the complications of the glandular follicles of the maternal placenta.

I have nothing to add to what I have written upon the completely developed placenta.

The addition or modification that is required does not, therefore, concern the fact itself, but my manner of judging it. I had believed that that which depends upon a neoformation of special histological elements was itself dependent upon the transformation of the folds of the preëxisting uterine mucous membrane.

I must now, with regard to the utricular glands, fill a vacancy left in my memoir.

Positive observations upon what becomes of these glands at the place corresponding to the placenta were altogether wanting. Even those who had believed that, at the beginning of pregnancy, at least, the villi of the chorion penetrate into the utricular glands, were forced to confess that they had never met any traces of these glands in the placenta.

Florinsky very recently affirmed the same thing, after having made the most minute researches in order to find them.[1]

Now, in following the mode of formation of the placenta in the dog, I have been able to establish a precise observation upon this point, an observation which led me to discover an error that I had committed, as I have pointed out above.

At the tenth day of gestation the dilatation of the utricular glands is already remarkable at the place

---

[1] Protocols des Vereins russischer Aerzte zu S. Petersbourg, page 141. 1863–64.

where the ovum was arrested. Their diameter exceeds by a third that of the same glands in the empty portions of the uterine horns. This dilatation must be very rapid. Between the sixth and the tenth day at the point where the placenta begins to form the glands show themselves so enlarged and distorted that they resemble enormous cavities. Their epithelium is in proliferation. Consequently, those who do not follow closely the transformation peculiar to these glands from the beginning to the end of pregnancy have great difficulty in recognizing them, and I was myself very far from the true inference while studying the fully-formed placenta.

On reading my memoir again, it will be seen how I was deceived as to the cavities that I saw.

In the dog, also, they are formed out of the utricular glands thus distorted, and I took them for the *culs-de-sac* of the large folds of the uterine mucous membrane, of which I believed that the maternal placenta was formed.

In Plate IX., *d, d,* we see these cavities, which are really, as I have just said, only the utricular glands, thus dilated and distorted from the first periods of gestation. Now, the plate needs no change as regards anatomical truth; it is the interpretation I had given of it, which needs the modifying and correction which I proceed to make.

In the first period of placental formation in the cat, the production of the phenomenon is readily followed. It is also seen how the connective tissue surrounding the utricular glands serves as a direct support to the special new tissue, formed of delicate, rounded cells, analogous to those I had described in the serotina of woman, and which compose the lamellæ that, as I have stated, are first lifted up and afterwards transformed in order to be reunited into glandular follicles.

The dilatation and rapid degeneration of the utricular glands lead us to believe that this phenomenon is produced in consequence of the occlusion of their orifices, occasioned by the prompt neoformation of the cells constituting the glandular or maternal portion of the placenta.

We ought, therefore, to consider these cells as representing the serotina in the human species.

I think it unnecessary to add that it is precisely the portion of the mucous membrane of the uterus where the transformation of the utricular glands takes place, which degenerates with them after delivery into a fatty substance, and disappears gradually, as I stated in the study upon the dog.

The cellular elements of new formation have no analogies in healthy adult organisms, but only in the tissue which Virchow calls mucous, and which abounds in embryo.

This would suffice, without taking into account remarks already made, to bring the mode of formation of the single placenta in certain animals near to that of the human placenta. In both cases alike, the glandular organ owes its origin to a neoformation of special cellular elements, which in woman takes the name of decidua serotina; its existence being denied in animals for want of exact observations.

But only after having observed the first periods of placental formation could this be positively asserted.

SECTION II.

THE HARE.

In order to be convinced that the special cellular tissue which, as I have said, is formed of the sub-epithelial connective tissue of the uterine mucous membrane in the cat is truly that which constitutes the serotina in woman, it is only necessary to study the placental structure of the hare.

In the hare (*lepus timidus*) the phenomenon which I am about to explain is much more clearly seen than in the placenta of the rabbit; and this difference between the placentæ of these two so closely-allied species is not the only one.

The placenta of the hare is developed by a thick layer of round cells, as voluminous as those of the serotina in woman.

The transverse sections of this raised layer of simply cellular tissue allow us to see in its interior numerous quite large cavities, various in form; the smallest are full of blood. This offers an example, or is rather the elementary form, of the large lacunæ of the human placenta seen in an animal. The larger cavities are filled with a somewhat dense fluid of a caseous appearance, especially at the bottom, that is to say, towards the uterine surface of the placenta. The presence of large cells in this fluid leads to the supposition that it is formed in the interior of the cavities by the deliquescence of the large cells of the thick serotina detached from the internal wall of these cavities.

Only towards the fœtal surface of the placenta of the hare, the glandular tubes into which the very short villi of the chorion penetrate are formed by the cells of the serotina, which the walls have furnished to the larger cavities. In this animal, the serpiginous character of the glandular tubes is only observed upon the most superficial layer of the foetal surface of the placenta.

It is then settled beyond doubt, that in animals also a serotina exists. In the hare it is even very thick. The form of the elements which compose it is identical with that of the elements of the human serotina.

To sum up, if the variation of development between the placentæ of animals and the human placenta is very great, the unity of the formative type of the glandular organ is only more clearly and surely demonstrated by this very difference.

SECTION III.

THE GUINEA-PIG.

Another difference between the placenta of the rabbit and that of the hare is more clearly seen in the guinea-pig, and a very important conclusion may be deduced from it.

In the hare, but much better in the rabbit and in the guinea-pig (*cavia cobaya*), when the placenta is completely formed, we see that the very numerous and complicated vascular loops which constitute its foetal portion are surrounded by a layer of large cells of the serotina, which have the appearance of large corpuscles of the connective tissue. This cellular layer presents the characters of the walls of the glandular follicle without in reality possessing all their anatomical parts.

Is then my statement regarding the structure of the maternal portion of the placenta invalidated by this observation in some few animals?

On the contrary, it only confirms the anatomo-physiological aphorism already admitted by science, namely, that secretory cells are sufficient to represent a glandular organ.

But there is more to be said upon this subject. In the human species itself, during the first periods of placental formation, the villi of the chorion are in immediate contact with the cells of the serotina.

Therefore the placenta of the hare, complete and at term, and yet more that of the guinea-pig and of the rabbit, shows in an easy and permanent way the transitory and incomplete first period of placental formation in woman; as the numerous vascular villi, observed by Bruch upon the uterine surface of the *squalidæ*, represent the first periods of formation of the placenta in the cotyledons of the cow.

Thus it becomes necessary to modify the explanation of Plate VII., Fig. 3, from that given.

What I had supposed to be the beginning of the neoformation of the glandular organ in the rabbit is in reality only the mass of the cells of the serotina, which lean against the villi without being transformed. The phenomenon in question is easily observed in the first stage of formation; but when the placenta is organized the enormous development and the very complicated ramifications of the vessels of the fœtal placenta constitute a kind of agglomeration of vessels, crowded and pressed close together, which renders the cellular layer of the serotina much more difficult of demonstration.

IV.

ON THE FORMATION OF THE PLACENTA IN WOMAN AND IN THE MONKEY.

SECTION I.

WOMAN.

IN the human species, better and more surely than elsewhere, I had been able to study the formation of the glandular organ by the cells of the decidua serotina.

The observations I have just described upon the development of the placenta in the cat, and upon the completely formed placenta of the hare, support my conclusions. Those I have since made directly upon the mode of formation of the human placenta will confirm them still further.

The proliferation of the characteristic cells of the serotina (Plate X., Fig. 1, f, f, and Fig. 3, b, b) is much more abundant in the uterine region, which corresponds to the place where the ovum has been arrested. A part of the chorion is thus brought into immediate and direct contact with the new-formed cells, that is to say, with the serotina. When the villi of the chorion begin to develop, even before their vascularization, they insinuate themselves, spread out, and ramify between the cells of the serotina. Hence in the first stages of formation of the human placenta, in vertical sections as well as in horizontal ones, we find under the microscope the self-same figure of villi cut across, which resemble perfectly round disks in immediate contact with the cells of the serotina, and entirely clothed by them.

Some of these villi are also cut in a longitudinal direction. In the latter, better than in the others, the external epithelium is clearly distinguished, which I had not been able to see with perfect certainty in the placenta at term.

It follows from these facts, therefore, that during the first periods of development of the human placenta, the epithelium of the chorial villi is, without any doubt, in immediate contact with the cells of the serotina only.

The vascularization of the maternal placenta takes place before the vascularization of the villi of the chorion, and before the complete formation of the foetal placenta. I have been able to pursue with ease the manner in which this is effected in two abortions, of one and of two months respectively. In the same way as the villi of the chorion penetrate into the foetal surface of the placenta between the cells of the serotina, so the vascular loops which spring from the uterine vessels insinuate themselves between the cells of the serotina through the uterine face of the placenta. The vascular loops are not distributed and do not ramify between the cells of the serotina; but during the early periods of gestation, they become diffused and dilated exactly like the capillary vessels in the erectile tissues.

These dilatations augment by degrees and form large projections which rise towards the foetal surface of the placenta; still increasing in size, they at length surround the chorial villi, which remain, however, covered with a layer of cells of the serotina. It is, in fact, these cells that are finally changed into a glandular organ (Plate X., Fig. 3).

In the first stages, during which the small lacunæ of the human placenta are formed, it is not difficult to ascertain the period when, together with the blood which they contain, the cells of the serotina are transformed into corpuscles of connective tissue, of which the fibrous walls of the same lacunæ are afterwards formed, just as we see them clearly in the completely developed human placenta (Plate X., Fig.1, h).

The transformation of the cells of the serotina into corpuscles of connective tissue, in proximity to the new lacunæ, which have an irregular shape, and in which it is not possible to distinguish the very thin wall of the dilated capillaries, might lead to the supposition that these lacunæ are the result of the rupture of the capillaries, or of the gradual expansion produced by the sanguineous effusion.

The observations of Robin and of Legros upon the dilatation of the capillaries in the erectile tissues do not permit this supposition; and it is altogether eliminated by the very easy demonstration which allows us to discover the epithelium of the circular sinus of the complete placenta, which forms a part of its lacunose system.

The very delicate wall of the capillaries thus dilated so enormously and strangely in the human placenta must of necessity line the external walls of the glandular organ which surrounds the villi and which is composed of the cells of the serotina. I will state, however, that the most attentive and direct observation has not yet enabled any one to distinguish this thin vascular membrane from the external membrane which covers the villi.

All that I can add to what I have written upon the

anatomical structure of the placenta has reference, as I have just indicated, to the first periods of its formation, that is to say, its histogenesis.

I review, therefore, what I have written upon this subject in the following manner.

In no period of the development of the embryo does the maternal blood come in contact with the villi of the chorion.

During the early period of development of the fœtus, in the majority of the mammals that I have examined, the vessels are as much lacking on the part of the fœtus as upon that of the mother.

In the cat and in woman the nutrition of the fœtus is carried on directly by an osmotic exchange between the cells of the serotina and the mucous tissue of the villi, and in woman by means also of the epithelium, which I have not hitherto been able to observe in the chorial laminæ of the cat.

When vascularization has taken place in the fœtal and in the maternal placenta, the osmotic processes are established between the fluid secreted by the cells of the glandular part of the mother and the vascular part of the fœtus.

In the human species, and in certain animals with a single placenta, the glandular organ owes its origin to the neoformation of a tissue of special cells, which in the human species has received the name of decidua serotina, the existence of which in other animals has, until the present day, been unjustly denied.

The placenta at term in the hare and in the guinea-pig presents the first transitory period of the formation of the human placenta.

When the cellular neoformation, or the serotina,

does not take place under the form of a layer of greater or less thickness, the maternal placenta is constituted by means of a special papillary neoformation, such as we observe during its different stages of development in the cotyledons of the cow.

In the elements of neoformation there is no change except in shape. In the cow, also, and in other animals, the papillary projections show themselves under a special form of elements, which we call serotina when it presents a uniform layer of cells.

This same mode of development may be reasonably supposed to exist in other ruminants, and even in some animals with a single placenta, as the mole, for example.

With all this, I do not pretend to affirm that the two modes of placental formation observed and described are the only ones which take place among mammals. They are the only ones which I have so far been able to determine, and they seem to me sufficient to establish the fact that there is always a true and actual neoformation.

The glandular organ which is the result of it departs during its development from all the known laws which govern the genesis of glandular organs in living beings; it cannot be confounded in any manner with the modifications of the mucous membrane, or of the preëxisting uterine glands.

SECTION II.

THE MONKEY.

I come in the last place to fill another blank that I was forced to leave in my memoir.

The conclusions which I had been able to make, up to that time, upon the anatomical structure of the completely formed placenta, not having been extended to its mode of development, had led me to believe that the differences between the human placenta and that of animals were much greater and more important than they now appear.

I then proposed a scientific question of positive fact, so to speak, and asked whether the placenta of the quadrumana would have the type of animals or that of the human species.

In the mean time, my good fortune and the courtesy of my two colleagues, Professors Bassi and Rivolta, of Turin, have enabled me to study a placenta at term of a *Cercopithecus sabæus*. Between the human placenta and that of this species of monkey I have met with no notable differences.

I can myself, therefore, answer the question which I had proposed in order that others might have the opportunity of resolving it. The type, and much more than the type, the anatomical structure, of the placenta of the monkey is identical with the placental structure of the human species.

MONOGRAPH

UPON

THE UNITY OF THE ANATOMICAL TYPE OF THE PLACENTA IN THE MAMMALIA AND IN THE HUMAN SPECIES, AND THE PHYSIOLOGICAL UNITY OF THE NUTRITION OF THE FŒTUS IN ALL THE VERTEBRATES.[1]

WITH reverence do I recall the honored memory of Filippo Ingrassia, who, in the classic land which was the cradle of the revival of letters, eulogized at the close of the sixteenth century the fruitful union of human and veterinary medical science in his famous oration, published in 1568, entitled "Quod Medicina Veterinaria formaliter una eademque sit, cum Nobiliore Hominis Medicina." At a period so remote from our own, and when veterinarians were only poor farriers, holding a very humble position compared with those skilled in human surgery, this was the happy intuition of a noble mind, which boldly outran the times, sowing the seed of that entire revolution which after three centuries comparative and experimental pathology have introduced into all medical instruction.

The work of Ingrassia in Italy was carried on by others, among whom were Lancisi,[2] Ramazzini,[3] Luigi, Galvani, and Gandolfi, who published at Bologna, in

[1] Translated from the Transactions of the Academy of Sciences of the Bologna Institute, 1877.

[2] Dissertatio Historica de Bovilla Peste. Romæ, 1715.

[3] De Contagiosa Epidemia quæ in Patavino Agro et tota ferc Veneta Ditione in Boves irrepsit. Padua, 1712.

1817, " Parallels between Human Maladies and those of Animals," and Pucinotti, whose work on " A Contagious Epizoöty and on Contagions in General" appeared in 1825. It is unnecessary here to mention the numerous works of Alessandrini. It was he who, founding in Italy, and with us, the first museum of comparative pathological anatomy, procured a true glory for our university, which the most learned followers of medical science in Europe commended and admired as the ripe fruit of the good seed sown in Italy by the great Sicilian physician.

This glory, all our own, was diminished not long since,[1] and the ancient edifice erected with so much care and affection, and supported by the self-sacrificing labors of Italy's most distinguished medical men, seems destined to immediate and certain ruin; but, happily, in the civilization of the present day, errors pass away and die with the men who taught them, while for science and its true followers the laborious conquests of human learning are never lost. Comparative pathology will make its way among those who cultivate medical science throughout the civilized world, and history will gladly record the past glory of our Athens, where for a century was taught, not in vain, with Ingrassia, " quod medicina veterinaria formaliter una eademque sit, cum nobiliore hominis medicina."

But not by cherishing the memory of a glorious past, and not with the weak accents of useless complaint, is the pain of the present removed. It is with

[1] By a royal decree, dated August 24, 1876, the Veterinary Institute was separated from the Medical Faculty in the Royal University of Bologna.

study and unwearied application that the future is prepared and slowly matured.

One of the most illustrious anatomists and histologists of learned Germany has lately asserted that the human placenta, regarded as a whole, is a formation very spongy in character, abounding in blood, which does not easily allow investigation with the ordinary methods that are most useful in all anatomical researches, namely, the knife and injections, — and he finds in this the reason for the uncertainty that exists with regard to its intimate structure. The illustrious German somewhat forgets the important precept left us from the old Italian anatomical school, which was the same inculcated by Ingrassia for pathology, that it is of great service in clearing up intricate and difficult questions of human anatomy and pathology to consult comparative anatomy and pathology. Kölliker, in discoursing of the human placenta, contented himself with noticing simply and incompletely, by way of supplementary instruction, some of the manifold differences met with by observers in studying the placenta of mammals.[1] He did not seek to fulfill the more serious and important duty for an anatomical professor, who, after observing and studying the multiplied differences of form which are presented in one organ in different animals, endeavors, by means of careful and minute comparisons, to explain more readily the physiological unity or the office of the organ.

Following the path traced out by our great masters, and the instructions of Malpighi, advantageously

---

[1] Kölliker: Entwickelungsgeschichte des Menchen und der höheren Thiere, B. 1, s. 331. Leipzig, 1876.

applied by the moderns to inquiries into the process of evolution in the organs and tissues, I believe that by the aid of comparative anatomy I have attained the end aimed at in my long-continued studies on the structure of the placenta. I can now show, with some force and clearness, that where the usual anatomical methods fail in ascertaining the structure of the human placenta, comparative anatomy alone is able to reveal it, and to demonstrate the unity of the anatomical type of the placenta under whatever manifold forms and varying external appearances it may be presented in the different classes of mammals and in the human species. This conclusion, which I have reached after many years of assiduous research, is not only important in itself, but is the more so because it goes beyond the limits of simple anatomy and invades the field of zoölogy. Learned men, like Owen, Huxley, and Kölliker, have placed as a basis of the fundamental distinctions in the mammalia the presence or absence of the placenta, or of facts connected with its intimate structure, which renders the path I now follow very difficult. If I shall be successful the merit will not belong to me, but to the university in which I was educated in medical knowledge, where my illustrious master learnedly explained the pregnant teaching left us by Ingrassia, and to my having pursued in the study of the human placenta the method adopted by Fabricius of Acquapendente [1] three centuries ago.

I shall divide my work into two distinct parts, since no better arrangement of the important subject I have undertaken presents itself to me.

[1] De Formato Fœtu. Padua, 1604.

In the first I shall indicate some new observations of facts which serve to explain the origin of the elements of the decidua and the maternal portion of the placenta, a point which hitherto has lacked a clear and positive demonstration. I shall notice some peculiarities in the rodents, heretofore unsuspected, relative to the complete destruction of the uterine mucous membrane and the subjacent parts before the formative process of the decidua takes place. I shall then describe the structure of the fully developed placenta in the *Cavia cobaya,* in which, besides notable peculiarities relative to the reflected decidua, we find in a single placenta the remarkable example of the form and structure seen in the multiple placenta of a ruminant united with that which is observed exclusively in cases where the placenta is single. In summing up these observations, I shall establish the fact of the destructive process not only of the uterine mucous membrane, but of the thick, glandulo-vascular layer beneath, leaving bare the inner muscular wall, —a process taking place beyond question in the rodents immediately after conception. Finally I shall endeavor to demonstrate that the same destructive process, differing only in degree, takes place in the uterus of all the mammalia, and is indispensable in all cases to the establishment and development of the neo-formative changes from which will result the maternal portion of the placenta.

These facts, hitherto imperfectly considered, or even entirely unknown, open the way for showing, in the second part, that, amidst the manifold and striking anatomical differences met with in the placenta in the different classes and species of mammals, there is always preserved an unfailing typical unity of anatomical structure, and that the variety and the multiplicity of forms depend only on a few very simple modifications observable in the two fundamental parts which constitute the placenta in the mammifera and in the human species.

This knowledge of the typical forms of the two constituent parts of the placenta wonderfully helps in judging of the villi, which are developed after conception in the gravid uterus of certain viviparous fish, and which represent the typical form of the maternal portion of the placenta in mammals, as a primordial manifestation of the placenta in vertebrates. By this link between the oviparous vertebrates and the mammalia is clearly shown, in all, the physiological unity, by the mode of nutrition of the fœtus, completing on this side the modern discoveries in embryology.

I.

The opinions which have hitherto been entertained concerning the origin of the cells which enter so largely into the formation of the maternal part of the placenta have been exceedingly vague and uncertain, when they were not altogether false, as were the teachings of those who asserted that the maternal portion of the placenta was merely a tumefaction or a transformation of the uterine mucous membrane. In time past I showed myself inclined to believe that these cellular elements might proceed from a transformation of the corpuscles of the sub-mucous connective tissue of the uterus, and from a proliferation of the newly-formed elements, but at the same time that I pointed out the facts which seemed to confirm this idea, I stated that so distinguished an anatomist as Waldeyer had informed me of a suspicion he entertained, to the effect that the new cellular elements originally constituting the different portions of the decidua, and then the maternal part of the placenta, might proceed from the walls of the uterine vessels. For a long time no positive observation was afforded me for holding either opinion as demonstrated, and, for my own part, I confess that the answer to such inquiry did not at first appear to me to have the deep scientific interest that really belongs to it.

Turner, in the first part of his last important work [1] on the placenta, although concisely affirming with Owen that without decidua there is no formation of placenta, does not touch the important question of the origin of the decidua; and Kölliker, a no less distinguished anatomist, confines himself to the remark [2] that "the decidua is a transformation of the uterine mucous membrane, and not a new membrane, or the product of an exudation, as was once believed," and as farther on he says [3] that "the different portions of the decidua have originally the same structure," the meaning seems to be that the origin of the decidua is to be found in the transformation of the preëxisting elements of the uterine mucous membrane, which, as we shall see, is very far from the truth, the decidua being due to a real neo-formative process, as I of late have endeavored to prove.

[1] Lectures on the Comparative Anatomy of the Placenta, page 113. Edinburgh, 1876.

[2] Entwickelungsgeschichte des Menschen und der Höheren Thiere, page 326. Leipzig, 1876.

[3] Op. cit., p. 336.

It is, however, well to notice here that this famous anatomist does not attempt to show which elements of the mucous membrane they are that compose the decidua, nor by what means they are transformed, although such an investigation should be of the highest interest for him, since the crypts or the simple and compound glandular follicles which I demonstrated to be of new formation, and to proceed from the previous neo-formation of the cells of the decidua, in the case of diffused and multiple placentæ (observations largely confirmed by Turner in the above-cited work), are with Kölliker only deepening depressions or tumefactions of the preëxisting uterine mucous membrane, formed during pregnancy and disappearing after delivery. The remote cause of these supposed tumefactions and hollows, as well as that of the transformation of the uterine mucous membrane for the development of the decidua, thus remained in the darkness of uncertainty.

I have elsewhere remarked that in the females of multiparous mammals having a single placenta, whether of zonarial or of discoidal form, the fecundated ova, when they have descended into the uterus, remain sequestered in the places where they were arrested, because in the intermediate segments of the uterus between ovum and ovum, the uterine mucous membrane was so tumefied that its longitudinal folds intersected each other in such a way as to produce a complete occlusion of all the segments of the uterus that remained empty.

I have met with this considerable tumefaction of the mucous membrane over the whole internal surface of the uterus of a dog in heat, and the turgescence of the mucous membrane was already such that the uterus being cut across, its internal cavity appeared like a stellate fissure, the tumefied folds of the mucous membrane being united with each other.

In this state of turgescence in the uterus of the dog in the state of heat which precedes conception, as well as when conception has taken place, no change is observed in the histological elements in the portions of the uterus where the ova have not been arrested. The increase of volume in the utricular glands which is very quickly initiated in the earliest periods of conception is not sufficient to cause the considerable enlargement in the part, which is evidently due almost exclusively to a strong vascular turgescence. This turgescence of the uterus, already observed by other anatomists at the commencement of gestation, has been very recently noticed by Turner in the uterus of the cat,[1] and it is the only known fact that serves to point out that in some cases prior to conception, in others at that period, all the internal surface of the uterus is prepared for those later changes which will be indispensable at the indeterminate point or points where the ovum or the ova are to be arrested after fecundation.

It is evident that the only way of ascertaining the origin of the cells of the decidua is to make the investigations on the gravid uterus at the first moment of conception; but this way had already been pursued with regard to animals having diffused and multiple placentæ, and with those also where the placenta is single, without obtaining facts so clear and evident as to warrant a positive opinion on this inter-

[1] Op. cit., p. 72.

esting subject, and it has been only very recently that I have succeeded in solving the difficult question by examining the gravid uterus of a rabbit at the precise moment of the beginning of the formation of the placenta.

I think the favorable time for this examination is very transient, and therefore offer the illustration of it in Plate I., Fig. 1. Fortunately, however, the observation is convincingly repeated in the examination of the placenta formation in the carnivora, but of this I shall treat subsequently.

The figure accurately represents the transverse section of a portion of the gravid uterus in the rabbit, about fifteen days after conception, at the place where the placenta begins to be formed. Here are at once seen three parts clearly distinct from each other, which correspond (*a, a*) to the muscular tissue of the uterus, on the inner surface of which the placental neoplasm projects (*b, c*). Above this is shown (*d, d*) the old uterine mucous membrane, tumefied, and about to be destroyed.

If we give an attentive examination to these parts, two very interesting facts are at once apparent; first, that the utero-placental vessels (*b, b*) have a lumen almost double that of the uterine vessels from which they proceed (*a′, a′*), and, moreover, that notwithstanding their greater volume, they show on their walls none of the anatomical characteristics which serve to distinguish the arterial from the venous vessels, and which are clearly visible in the uterine vessels. The second fact is that these vessels, instead of the ordinary walls, are surrounded with a uniform envelope of cells of a special character, which are exactly those found constituting the cells of the decidua serotina and the maternal placental tissue (c, c).

At first sight the vessels of the vascular net-work, clothed with a layer of special cells which stand out from the uterine muscular tissue and represent the placenta in this first phase of development, might be supposed to be the vessels of the vascular net-work of the old mucous membrane that had undergone transformation, and which I have already described as remarkably tumefied in the uterus of the dog even before conception has taken place. But this suspicion is removed by the special characteristics before pointed out in the walls of the vessels, — a fact on which I shall hereafter have occasion to dwell in order to show that it is the endothelium alone that forms their walls. A stronger argument, however, than any minute observation, is the exceptional fact that all the old uterine mucous membrane, including the crypts that represent the utricular glands in this animal, and the underlying connective layer with its vessels and nerves, is already observed at this stage of development of the placenta, greatly changed, and in course of complete destruction, detached from the whole internal surface of the muscular tissue of the segment of the uterus where the ovum has been arrested, and above the placenta at the point where that is formed (Plate I., Fig. 1, d, d). This separation and great destruction of the uterine mucous membrane and of the connective layer beneath, which I shall examine more minutely in other rodents, I have pointed out and represented in the rabbit, because they show, beyond all doubt, that the placenta in this animal cannot be produced by the transformation of the anatomical elements of the mucous membrane existing before the act of conception, and that the utero-placental vessels as well as the uniform layer of cells which surround them are owing to a real neo-formative process.

In all animals having a single placenta that have hitherto been examined, the decidua vera and the decidua reflexa are alike due to a neo-formative process. The decidua vera, when detached from the inner wall of the uterus, consists of the elements of new formation, which have been arrested in their development, and those of the uterine mucous membrane, which were superimposed upon the new-formed cells. According to this view, in the rabbit the decidua vera and, what is of more importance, the decidua reflexa are both wanting, being represented only by the old uterine mucous membrane and the connective layer beneath, wholly detached from the inner surface of the uterine muscular tissue, and to be seen later in course of complete destruction between the external wall of the ovum and the internal wall of the uterus. We shall presently obtain more noteworthy examples in other rodents of the destructive process so extensively sustained by the mucous membrane and the subjacent parts, as far as the uterine muscular tissue, immediately after conception, and of the neo-formative process of the decidua vera and reflexa. These conclusions, demonstrated by observations on the gravid uterus of the rabbit, confirm my previous statement, that the vessels themselves are of new formation, and from them are developed the cells of the serotina constituting the maternal portion of the placenta, and exclude by a very evident fact any participation whatever of elements preëxisting in the uterine mucous membrane in the formation of the placenta.

The opinion, or rather the suspicion, thrown out by Waldeyer on the origin of the cells of the serotina was approximately true, but not the actual truth, since it is not from the preëxisting vessels of the uterine mucous membrane that these cells arise, but the vessels themselves are of new formation and of special structure, and therefore constitute an integral part of the placental neoplasm. I have above indicated that the origin of the cells of the serotina and placenta from vessels of new formation can be understood without difficulty by examining the structure of the placenta, even when fully developed, in the carnivora, which I shall now briefly consider.

Carefully studying the observations instituted by Turner on the anatomical structure of the placenta of the dog, the cat, and the fox, as well as that of the seal and the hyrax, all of which have a common type, I was glad to find that so illustrious an anatomist has confirmed my first observations on the placenta of the cat and the dog, namely, that the villi of the chorion do not enter into the utricular glands to form the placenta, as Sharpey and Bischoff had taught;[1] that as these utricular glands do not open into the cotyledons of the gravid uterus in the cow, as Bischoff, Eschricht, and Spiegelberg had admitted,[2] so these glands do not open into the placental crypts in the cat;[3] and that the crypts which are formed in the first period of gestation in the placental region of the carnivora are not owing to a simple enlarging of the

[1] Turner, op. cit., p. 72. [2] Op. cit., p. 69. [3] Op. cit., p. 74.

mouths of the glands, but are really of new formation.[1]

But I cannot agree with Turner that the aforesaid crypts, or, as I called them, glandular follicles, are produced in the dog and the cat, as well as in the sow and the mare, by an excessive growth of the inter-glandular part of the mucous membrane which is folded over to form them.[2] Nor can I admit that the morphological elements of the placenta in the dog, the fox, the cat, and the seal are indubitably like the crypts of the mucous membrane in the mare, a cetacean, or any other animal having a diffused placenta.[3]

Upon examining and comparing the descriptions given by Turner, and his accurate figures of the injected placenta of the fox,[4] with my own preparations and statements on the placenta of the dog and the cat, I at once perceived that, if the observations of Turner were exact, I had fallen into a serious error in my opinion of the intimate structure of the placenta in the carnivora that I had examined, having mistaken the utero-placental vessels of the maternal portion for the proper vessels of the foetal portion, and, in consequence of this error, having regarded as closed and new-formed glandular follicles, or glandular crypts of new formation as Turner considered them, parts which in their form have nothing in common with the crypts or glandular follicles properly so called.

All those who have attempted to study the intimate structure of the placenta, and know the serious difficulties encountered at every step in this pursuit,

---

[1] Op. cit., p. 75.
[2] Op. cit., p. 75.
[3] Op. cit., p. 111.
[4] Op. cit., Plate I., Fig. 1, 4, 5.

will readily understand that, not having previously injected the placentæ of the carnivora, the mistake I made was not only very easy but almost inevitable. Nevertheless, I do not seek to excuse myself by showing how one may be drawn into error, nor to shield myself with the illustrious name and teaching of Professor Turner, who, even after performing injections, fell into the same error, and received as true, as I have pointed out, my first incorrect conclusions. Enough for me, that, stimulated by the observations of Turner, and after repeating them, I have been able to correct my mistake, and through cognizance of it I have at length come to understand that, though in certain animals the glandular structure of the maternal portion of the placenta clearly appears under the aspect of crypts or common glandular follicles, as when it has the diffused or multiple form, on the other hand, when the placenta is single, whether zonarial or discoidal, the common glandular form is completely disguised, although without losing the fundamental character of a secretory organ. It was by repeating investigations in this direction that the idea was finally reached of the single anatomical type of the placenta in all mammals. But before coming to the conclusions let us examine the facts.

In the dog and the cat (and, as I have elsewhere shown, the same thing occurs in the human species also) the subjacent utricular glands at the place where the placenta is formed are dilated and altered in shape so as to produce a sort of trabecula with large meshes, or a spongy tissue, which is found between the surface of the inner muscular tissue of the uterus and the uterine surface of the placenta. There is a slight indication of this spongy condition seen in Plate I., Fig. 2, *a, a*, which shows a section of a placenta of a dog at term, with the vessels of the maternal portion injected.

The placental vessels have a uniform diameter, and form a net-work of not very large meshes, which reaches as far as the chorion, with which it is firmly joined (*b, b*). In the chorion (*c*) are seen, cut across, the fœtal vessels which, uniting, go to form the umbilical cord.

Every ramification of the net-work of the injected placental vessels is in direct relation with the uterine vessels, and everywhere surrounded by a uniform and rather thick layer of cells (*b', b'*); therefore all the maternal portion of the placenta in this animal is formed from the vascular net-work and the surrounding cellular envelope. We have, in short, in a more complex and permanent manner, the exact repetition of the fact transiently observed in examining the placenta of the rabbit at the beginning of its development, and represented in Plate I., Fig. 1. In some placentæ, then, even at the close of gestation, we obtain positive proof that the cells of the serotina and placenta originate from the walls of the placental vessels, which, in this case also, lacking the proper anatomical characteristics of ordinary vessels, are to be held as being themselves of new formation and special even by their exterior cellular covering.

Again, by injecting the fœtal vessels we have a clear and positive demonstration of the relations established between the fœtal and the maternal portions. The injection of these vessels affords convincing proof that the arterial, as well as the venous vessels, which have a very considerable diameter in the chorion that adheres to the placenta (c) immediately below this, form a thin, close net-work of capillary vessels in contact with, and branching out in the cellular covering that clothes the maternal vessels. This relation is shown in Plate I., Fig. 3, representing a horizontal section of the placenta of a dog, injected and highly magnified.

Not only is the minute capillary net-work of the fœtal vessels distributed among the cells which cover the large-meshed, maternal, vascular net-work, but the vessels also ramify among the cells of this envelope, as is fully shown at the points where the maternal vessel was cut in the section (g,g). This figure perfectly corresponds with that given by Turner of the placenta of the fox.[1] Therefore the idea of true glandular follicles, or crypts, as Turner calls them, identical with those in the cetacea, the solipeds, and the ruminants, is no longer in any way tenable; but of this question I shall treat more minutely in the second division of this subject. At present it suffices that I have represented the plan of the placenta in the dog, in Plate V., Figs. 11 and 12, in order that every one may judge of the correctness of the idea indicated in the diagram by a comparison with the exact natural figures which I have explained, for it is from such facts that one arrives at a conception of the single anatomical type of placenta in the mammalia.

The separation and destruction of all the uterine mucous membrane and the subjacent connective layer in the segments of the uterine horn of the rabbit,

[1] Op. cit., Plate I., Fig. 5.

where the fecundated ova were arrested, and the knowledge gained from this fact, that new vessels with walls entirely special are formed in the placenta, upon the external surface of which the cells of the serotina proliferate from the very beginning of its development, have allowed us to recognize this same fact in the placenta of the dog at term. But this knowledge is not always acquired with so much certainty, or when it can be obtained it is only during so limited a period, as we have seen, in the rabbit that the origin of the cells of the serotina in many cases remains uncertain because the observation was not instituted at the right time. Finally, there are not wanting certain cases in which we can in no way decide whether we are dealing with a true neo-formation of which the origin is unknown, or, on the other hand, with a simple transformation of the preëxisting anatomical elements of the uterine mucous membrane.

Careful examination and study of different animals at different periods of gestation can alone remove the many uncertainties met with in regard to this; and from neglect of this sole way of arriving at truth, able anatomists have been silent as to the origin of the cells of the decidua, or, what is worse, have considered as a tumefaction, or a transformation of the uterine mucous membrane, the development of the placenta, which is a true and actual neoplasia.

The better to prove these assertions of mine, I will now present the observations which I have instituted upon the changes which the mucous membrane of the uterus undergoes in the first periods of gestation in the rat, because they confirm not only that, in these animals, the whole uterine mucous membrane is not tumefied or transformed, but also that it is completely destroyed after conception in the place where the ova are arrested. Furthermore, in the rat there takes place a complicated neo-formative process which shows clearly enough that the new-formed anatomical elements are identical in the decidua vera as well as in the decidua reflexa and serotina, and that consequently it depends upon the point where the ovum attaches itself whether the decidua becomes serotina, reflexa, or vera.

The process of destruction in the old, tumefied mucous membrane at the place where the ova have been arrested is very rapid in the rat, of which I obtained positive proof by examination of the gravid uterus of a *Mus musculus* and of a *Mus decumanus* a few days after conception. In Plate II., Fig. 1, is given the transverse and complete section of a uterine segment of a *Mus decumanus*, in which the ovum had certainly been just arrested, as it had not yet completed the phases of the segmentation (Fig. 1, c, and Fig. 2, e); and it is interesting to notice that already, from the very first moments of gestation, the ovum is everywhere surrounded with a cellulo-vascular neoplasm which completely isolates it at the place where it was arrested (Fig. 1, f, f, and Fig. 2, c).

There were but few relics of the old mucous membrane in the segment of the uterus in question, namely, the remains of some portions of utricular glands not yet destroyed, near the inner muscular layer of the uterus (Fig. 1, g, g, and Fig. 2, d, d), and in the presence of a small cavity of triangular shape, at an internal point opposite the place where the ovum was arrested (Fig. 1, *d*), which is still covered with the old epithelium of the uterine mucous membrane. This cavity for a time represents the cavity of the horn before pregnancy, but with the progress of the development of the ovum it entirely disappears, and it is exactly below the concave base of this triangular opening (Fig. 1, *h*) that in the rat the placenta is always developed. At the period of gestation now under examination is seen a slight fissure extending from this cavity as far as the ovum (Fig. 1, *e*). This has been determined by the turgescence and transformation that have taken place in the uterine mucous membrane. At the edges of this fissure, and towards the place where we now find the ovum (Fig. 1, *e*), the epithelium of the old mucous membrane is already lost, but everywhere, whether the epithelium remains or has disappeared, instead of the old anatomical elements of the former mucous membrane there is observed a formless mass of cellular elements, amongst which winds a minute, vascular network (Fig. 1, *i, i*, and Fig. 2, *c*). To give an exact idea of the changes wrought in the uterine mucous membrane at the places where the ova were arrested, and in those which remained empty, I have given in Plate II., Fig. 2, a longitudinal section of the same uterus from which the preceding preparation was taken. In the middle is clearly seen the ovum (*e*) surrounded by the cellulo-vascular neoplasm (*c*). In the segments of the uterus which remained empty the external and muscular walls of the uterus are indicated at *b, b*, and the cavity of the uterus at *f, f*. It is easy to observe, as shown at *g, g*, the tumefied uterine mucous membrane with its utricular glands intact. In this section there is no longer any trace of the triangular cavity, nor of the fissure just now referred to in Plate I., Fig. 1, *d*, which also proves the rapidity with which certain changes occur, since the two preparations were taken from the uterus of the same animal.

The origin of the wall of the vessels of new formation in the cellular mass (*c*), which imprisons the ovum at the place where it has been arrested, cannot in the rat be positively ascertained in this first period of gestation, and were it not for the plain demonstration obtained by examining the gravid uterus of the rabbit, the first idea suggested by the study of the preparations is, that the new-formed mass proceeds from a complete transformation of the preëxisting anatomical elements. But the proof that in the rat, also, all the decidua proceeds from a cellulo-vascular neoplasm, of which there is no trace in the non-gravid uterus, is clearly seen by examining a later phase of development of the decidua and the placenta in these animals, as represented in Plate III.

I wish here to state that all the preparations and drawings for this work have been very ably made by my excellent assistant, Dr. G. Pietro Piano.

In Plate III. is shown half of a complete transverse section of a segment of the gravid uterus of a *Mus musculus*, in which the placenta has not yet reached the last stage of development.

The decidua vera at this period of formation is already detached from the uterus through the whole extent marked *f* and *g*. The portion *g* corresponds to that part of the primitive decidua which at the beginning of conception we observed nearer to the ovum, the placenta in the rat being always formed where we remarked (Plate II., Fig. 1, *h*) the base of that narrow triangular canal which still preserved the old epithelium of the uterine mucous membrane, and which represents the former cavity of the uterus.

Upon the inner surface of the muscular tissue of the uterus (Plate III., *b, b*), corresponding to the place whence the decidua vera has been detached, there is already formed a new epithelial layer which alone, at this period of gestation, contains all the future elements that constitute the uterine mucous membrane in these animals, as well as the thick, vasculo-glandular layer below it which is destroyed immediately after conception.

The decidua reflexa (*c*) is seen to be formed of two clearly distinct layers, — the outer (*d*), which is in relation with the inner surface of the uterine muscular tissue, being composed of cells analogous to those which we found to constitute the uniform mass of the cells of the decidua soon after impregnation, and the inner layer (*e*), which appears to consist of a sort of trabecula composed of enormous stellate cells; the spaces interposed between these cells are no other than the lumen of large vessels cut across. In the portion of decidua vera (*f*) continuous with this the external layer of the primitive cells has entirely disappeared, and there remains only the cellulo-vascular trabecula, the structure of which is also lost in all the remaining part of the decidua vera (*g*). It is to be noticed that a species of very imperfect pavement epithelium (*o, o*) covers not only all the internal surface of the different portions of the decidua, but likewise extends over the whole foetal surface of the placenta as far as the place where the chorion adheres to it (*m*). If it had not been demonstrated that the old epithelium of the uterine mucous membrane disappeared completely from the earliest period of pregnancy, the aforesaid epithelial cover of the decidua vera might be regarded as a permanent residue of this last, but the observations show that such an opinion is not to be maintained, and that here, too, is new formation.

In the portion of decidua that has become serotina (*h*) there is seen at this period of gestation the cellulo-vascular trabeculæ of the primitive decidua, like a layer between the two clearly distinct parts in the placenta of these animals, the uterine, which I have called glandular because destitute of fœtal vessels (*i*), and the fœtal, in which alone are distributed the vessels of the cord (*l*). The large stellate cells which surround the ectasic capillary net-work in this portion of the decidua serotina disappear altogether with the progress of development, and the two parts of the placenta are no longer to be distinguished except by the presence or absence of the fœtal vessels among the placental cells that clothe the maternal vessels.

In the period of pregnancy now under consideration, the chorion (*n*) adheres to the placenta only at its central part (*m*), the further blending of the chorion with the margins of the placenta not occurring until later, and then precisely at the place where, as I have pointed out, the decidua reflexa is observed (*c*). Thus is formed that chorial bursa which hermetically closes the whole fœtal surface of the fully-developed placenta. When I before described the placenta of the rat, I mentioned the gigantic stellate cells found at this place, and I can now positively state that these represent the old elements of the original decidua vera in the place where it has become reflected. Observations instituted on the complete destruction of the uterine mucous membrane and all the subjacent parts as far as the uterine muscular tissue, acquaintance with the elements, characteristic in size and shape, which constitute the primitive decidua, and the changes sustained by these peculiar elements in the place where the placenta is formed and elsewhere during the period of gestation, plainly prove that in the rat, also, it is not the old elements of the old mucous membrane that are tumefied and transformed, but that the decidua and the placenta are alike owing to a true and actual neo-formative process.

Remarkable, and of no little interest, are the anatomical peculiarities which the study of the formed placenta of the *Cavia cobaya* presents. I did not have a favorable opportunity for following step by step its earliest phases of development as I was able to do in the rat. But soon after the descent of the ovum into the uterus I observed with much certainty, in the uterine mucous membrane, the same facts which I have enlarged upon in the discussion of the uterine mucous membrane of the rat, in the segment of the uterus where the ovum is arrested. At this period the destruction of the mucous membrane corresponds perfectly with that I have described and represented in Plate II., Figs. 1 and 2. Observations are wanting with regard to the formation of the primitive decidua. Plate IV., Fig. 1, shows half of a complete transverse section of the uterus and placenta of the cavia fully developed. The decidua vera (c) appears composed of anatomical elements already altered, which give no suggestion of their previous form. That the old mucous membrane and the underlying vasculo-glandular layer are completely destroyed in this animal, also, we have ample proof from the observation I have mentioned, from the thorough change in the mucous membrane early in the conception, and from the fact now known that the inner muscular layer of the uterus (a, a) is found to be covered by a simple epithelial layer (b), precisely as we have seen in the rat, which itself alone represents all the parts that have perished. A noticeable peculiarity seen in the decidua vera of this animal belongs to the epithelial covering with which it is furnished, as well on its uterine surface (c') as on its foetal or inner surface (c'').

The external surface of the decidua vera already completely detached from the uterus appears covered with an epithelial layer, but it cannot be decided whether this is of new formation, or is only the outermost layer of the new epithelium, which was developed on the uterine muscular tissue (b) that remains adhering to it in the act of separation. In the rat we have obtained only indirect evidence for the assertion that the epithelial layer that covers the inner surface of the decidua was itself of new formation; in the guinea-pig we have positive proof, because we see it in direct continuation with the epithelium of the decidua reflexa (h), which in this animal has a special organization, and clothes not only the peduncle, but the whole of the placenta, even when it has reached its complete development (f, f).

In its general form the placenta of the cavia is pedunculate, and except in the central portion of the fœtal surface, which adheres to the chorion (*l*), is all covered, as I have said, by the decidua reflexa, which presents special and differing characteristics in the part which invests the peduncle and in that which covers the surface of the placenta.

The decidua reflexa all around the peduncle of the placenta (*e, e*) has the appearance of a large fringed fold, and is, on the exterior, covered with an epithelial layer which might be thought a festooned folding of the old hypertrophic mucous membrane. That it is really owing to a neo-formation is convincingly shown by the quality of the rounded and special cellular elements observed in its interior (*n*), in continuation with those which form the peduncle of the placenta, and which are entirely different from the ordinary corpuscles of connective tissue. It is proved further by the entire absence of the utricular glands, and by the special characteristics of the large utero-placental vessels which traverse its external portion (*e*), in which are none of the anatomical elements that serve to distinguish the arteries from the veins, though in their normal state their diameter is far less. In the place, then, where this portion of the decidua reflexa is carried over upon the uterine muscular tissue (*d*) the proof of the neo-formation is better shown by an interlacing of utero-placental vessels, which in their transverse sections are seen everywhere surrounded by cells of new formation. This anatomical disposition, which is observed in this animal at one point only of the decidua reflexa when fully developed, corresponds perfectly with the facts which I have pointed out in the study of the formative process of the placenta in the rabbit, where I obtained positive proof that the cells of the serotina originate from the walls of vessels, themselves of new formation. In a limited portion, therefore, of the decidua reflexa in the cavia, and also when the placenta has completed its development, we obtain confirmatory evidence as to the origin of the serotinal or placental cells; nor is it only the decidua reflexa surrounding the peduncle of the placenta that presents characteristics important enough to exclude all suspicion that we are not dealing with neo-formed elements, for the same conviction follows an examination of that part of the decidua reflexa which nearly envelops the body of the placenta. This special external covering of the placenta furnished by the decidua reflexa, which I have observed only in the guinea-pig, is formed from the large cells of the serotina. Although there is evident trace of these cells at the point where the decidua reflexa of the peduncle is borne upon the body of the placenta, their volume gradually diminishes, until this covering of the placenta having reached the place where the fœtal vessels penetrate and the chorion adheres to the placenta, these cells and the firm neoplasm that surrounds them are blended with the connective elements of the chorion (g). On the placental side this layer of decidua reflexa covers a sort of superficial and vascular sinus, and on its external face is lined with an epithelial layer of irregular thickness (i, i), on which are here and there epithelial prominences rising, like buds of various shapes and sizes, from the surface of the placenta.

This singular epithelium is in direct continuity with the uniform epithelial layer that envelops the decidua reflexa in the peduncle of the placenta, and with that which clothes the inner surface of the decidua vera; therefore the neo-formation of this, too, seems to me indisputable.

The central part of the peduncle of the placenta (n) consists of a loose, delicate, and very peculiar tissue formed of cellular elements, varying in size (n') from gigantic cells to small round nuclei, irregularly distributed in a mass of protoplasm of gelatinous appearance. In the midst of this singular tissue, which holds the place of the old elements of the uterine mucous membrane, is observed a net-work of capillary vessels, irregularly ectasic, from which are formed the utero-placental vessels. Eschricht had already remarked the dilatations of the maternal vascular loops in the placenta of certain animals, and Turner finally confirms this observation in the placenta of the cat, the dog, and the fox. With much acumen he considers such dilatations as a first intimation of the lacunose circulation witnessed in the placenta in the quadrumana and in the human species.[1] The ectasia which I have now noted in the small vessels of the peduncle of the guinea-pig corroborates the opinions entertained by Eschricht and Turner. We shall see the importance of this fact later.

The anatomical structure of the placenta in the cavia is most singular, since although, as in the rat, it may be clearly divided into two parts, yet that portion which I have called glandular in the rat has in the cavia the exact form of a cotyledon of the cow

[1] Turner, op. cit., p. 85.

(*o*); I shall therefore call it cotyledonal, reserving to the other the name of placental, properly so called, because the relation of the fœtal portion with the maternal takes place in this in the manner which we have observed in the placenta of the carnivora (*p*).

To form an exact idea of the structure of the placenta in the cavia, we may imagine it an empty tunnel, closed at the lower, narrower part. The upper annular and peripheral part of the tunnel (*p*) is formed from the placental portion, the lower part, the *cul-de-sac*, from the cotyledonal portion (*o*); the internal cavity of the tunnel is filled with the tissue and vessels of the chorion (*g*).

The cotyledonal portion (*o*) rises from that singular tissue I have spoken of as forming the peduncle of the placenta. It has the shape of a cup, the concave part of which presents a surface elegantly undulated and festooned. This surface consists of a continuous layer, not of uniform thickness, of similar round cells set rather close together, and, at the place where the wave-like folds rise, there is a rich network of small, irregularly ectasic vessels. All the concave, undulate surface of this cotyledonal portion is completely filled with chorial tissue (*g*), in the midst of which the fœtal vessels ramify. The identity in form of this portion of the placenta in the guinea-pig with that of a small cotyledon in the cow would be complete if the chorial elements did not adhere to the surface of the undulate layer of the cells which I have just pointed out, and which are no other than cells of the serotina disposed as are the placental cells of a vaccine cotyledon.

In the placenta of the cow both the fœtal and maternal portions have a proper epithelium, and are separate from each other, while in the cavia the epithelium is wanting in both, and they are joined together. To give an exact idea of the anatomical structure of this cotyledonal portion of the placenta of the cavia, I represent a small section of it, highly magnified (500 diameters), in Plate IV., Fig. 3. At b is indicated the layer of cells that line the cavity of the cotyledon, at a the net-work of the maternal ectasic capillaries, from the walls of which is elaborated the before mentioned undulate cellular layer, at c and d the chorion and its vessels, to show clearly the relation existing in this place between the fœtal and the maternal portions.

The not very thick layer of cells over the whole of the cotyledonal portion is remarkably expanded upon all its upper edge, and constitutes the thick ring, as I have said, of the tunnel (Plate IV., Fig. 1, p). This placental portion is formed by a minute web of fœtal and maternal vessels; these last appearing of a black color because they were injected in the preparation. The maternal vessels are everywhere surrounded by a somewhat thick layer of cells, rather larger than those I have described in the cotyledonal portion. In the midst of these cells, which are so close together as to form almost a compact mass, traverse the fœtal vessels, which, in the figure, are seen empty and of a white color. The chorial tissue (g) is still adherent, in this inner and central part, to the placenta, as we have seen it attached to the surface of the cotyledonal portion, and over the whole fœtal surface of the placenta (l). I have not been able to ascertain with certainty that the vessels of the central part of the chorion, which are spread out in the cavity of the cotyledonal portion, communicate with the fœtal vessels that traverse the placental portion; rather has it seemed to me that the refluent blood of these two parts of the placenta mingles only in the cord before being carried to the fœtus.

In the rat, and more clearly in the cavia, appears the interesting fact of a two-fold anatomical constitution in the placenta, which allows us to infer that in these animals the double office of the placenta, that of providing for the nutrition, and the respiration of the fœtus, may be carried on separately in each part.

In Plate IV., Fig. 2, is represented, highly magnified (500 diameters), a section of the aforesaid vascular portion of the placenta of the cavia, as here are best seen the relations of its two fundamental parts. At a are indicated the injected maternal vessels, everywhere surrounded by a layer of cells (b). The empty fœtal vessels (c) course in the midst of these, and it is only their thick trunks, at some points surrounded by abundant elements of the chorion (e), that came in contact with the elaborated placental cells surrounding the maternal vessels.

The minute researches, which I have now explained, in the anatomical structure and formation of the decidua and the placenta in certain rodents have clearly shown some facts which had hitherto escaped the investigations of the most careful observers, and which are of no small interest, because they suffice of themselves to prove beyond all question the error of two different classes of anatomists, namely, of those who maintained that the placenta is formed by the entrance of the villi of the chorion into the utricular glands, and of those who taught, as has lately been the case, that it is developed by a tumefaction or some transformation of the elements preëxisting in the mucous membrane of the non-gravid uterus. But besides the complete elimination of the preceding errors, the facts demonstrated show with equal clearness that the decidua and the placenta are owing to a true and actual neo-formative process, so that putting my previous observations on this fundamental fact with those of even greater force now presented, the following conclusions appear to me warranted:

(1.) That in many animals and in the human species it is not merely the epithelium of the uterine mucous membrane, and perhaps a subtile, sub-epithelial layer, which is detached and destroyed after conception, to give place to the neo-formative process that produces the decidua and the placenta, but in certain animals (as the rodents), in the segments of the horns of the uterus where the ova were arrested, it is the whole uterine mucous membrane, including the underlying vasculo-glandular layer, which is subjected to a uniform and complete destruction.

(2.) That when there occurs so considerable a destructive process of the uterine mucous membrane, and of all the subjacent parts as far as the muscular tissue, it is not established in a uniform way. It may be called simple when, as we saw in the rabbit, the process is limited to the entire separation and complete successive destruction of all the parts above mentioned. The same changes in these parts may be called compound when a neo-formative process is mingled with the destructive, as we have seen in the rat from the very beginning of conception.

(3.) When the destructive process is simple, so too is the neo-formative process of the placenta which has its point of origin from vessels of new formation that issue from a portion of the uterine muscular tissue that is left denuded. When, on the other hand, it is compound, as has been observed in the rat, the neo-formative process of the placenta is clearly distinguished into two phases. In the first is formed the uterine decidua which lines all the inner surface of the segment of the uterus where an ovum has been arrested. In the second the neo-formative process becomes active at one point only of the inner surface of the uterus, and it is there that the placenta is developed. The neo-formation of the first phase, so plainly recognizable in the rat by the shape and size of the cells, is what constitutes the decidua vera.

(4.) In the place where the placenta is formed, the decidua is denominated serotina and reflexa. The peculiar structure of the decidua in the rat has shown us the existence of the serotina even at a somewhat advanced stage of placental development, and forming, as it were, a line of demarkation between the glandular and vascular portions so distinct in the placenta of these animals. In complete development every trace of the cells of the decidua serotina is lost in the interior of the placenta, and remains only at the edge where the decidua is styled reflexa. In the rabbit the neo-formative process of the placenta is much more simple, and is primarily established only at one point immediately below the old detached uterine mucous membrane. Of this we have positive proof by observation of the mucous membrane with its crypts and the subjacent connective tissue as far as the muscular wall in course of destruction above the placental neoplasm. These detached and altered parts form the deciduæ vera and reflexa; there is thus wanting in the rabbit that first phase of the neo-formative process which we have so plainly seen in the rat, constituting the decidua vera, the serotina, and the reflexa. In the place where the placenta is formed, it may be thought that the development of the decidua serotina and the reflexa is the same in the neo-formative process of the placenta, but in the structure of the decidua vera the differences are very considerable between the two species of rodents; the decidua vera in the rat resulting from special cellulo-vascular elements entirely different from the elements of the old mucous membrane, and from those of new formation accompanying the destructive process of the mucous membrane, whilst in the rabbit the new-formed elements constituting the decidua vera are altogether wanting.

(5.) The placenta of the *Cavia cobaya*, like that of the rat, is plainly separable into two parts, but in the cavia that part which I have called glandular in the rat has a better defined form, analogous to the structure of a vaccine cotyledon, thus presenting, united in a single placenta, two unlike forms of this organ; that of a ruminant in its central part, everywhere encompassed by that lace-work of fœtal and maternal vessels found throughout the whole extent of the single placentæ, whether of zonarial or discoidal form. This anatomical peculiarity, so well marked in the cavia and observed, although less perfectly, in the rat, may serve to explain the two-fold nutritive and respiratory office of the placenta in the mammalia.

Besides this remarkable peculiarity, the placenta of the cavia presents another, relating to the decidua reflexa, which has, in this animal, a special importance, throughout the whole period of gestation, in the anatomical constitution of the placenta. The new-formed elements of the reflexa, in addition to the modifications and the important offices they sustain in the peduncle of the placenta, constitute also its external wall, maintaining a peculiar structure up to the close of pregnancy.

Very recent observations have proved that the neo-formative process which produces the maternal portion of the placenta has its origin in vessels which, although in continuity with the uterine vessels, yet even in the placental neoplasm present altogether special characteristics, — either because the arterial as well as the venous vessels lose their external walls and thus the anatomical peculiarities which serve to distinguish them, and are formed by the endothelium alone (as Kölliker has indeed observed in the human species), or because from their external surface are generated the secretory cellular elements which have their proper characteristics and are integral parts of the maternal or secretory portion of the placenta.

The origin of these cellular elements has always been uncertain, but we have obtained its positive demonstration by examining the placenta of the rabbit during its first stages of development, and the uterine portion of the decidua reflexa in the cavia when the placenta has attained its complete development. With the aid of injections this important observation can be readily made in the formed placenta of the dog, which is identical with that of the fox, so well described and illustrated by Turner.

I have until now followed the old method of collecting new facts and particulars relative to the manifold initial forms of the placenta, as shown in the different species of mammiferous animals, but I think the time has arrived for leaving the analysis, and for attempting the anatomical synthesis of the placental organ.

The conclusions which I have presented relative to the complete destruction of the uterine mucous membrane, including all the underlying vasculo-glandular layer, at once appear in open and marked contradiction to facts positively ascertained, which show that the utricular glands of the non-gravid uterus not only remain but certainly increase in volume during the whole period of pregnancy. At first sight the contradiction is so plain and striking that one is inclined to believe that the production of the decidua in cases of diffused, multiple, or single placenta, when the old uterine glands remain and increase in volume, must be necessarily owing to an anatomical process entirely different from that observed in certain rodents, in which, from the very beginning of conception, the whole glandular layer is completely altered and destroyed.

But I do not think I err in asserting that the differences indicated, which seem so important, are only differences in degree, and therefore more apparent than real, and that even the two extremes which these facts include, namely, the destruction of the epithelium only of the uterine mucous membrane after conception, as takes place in the human species, and the destruction of a thick layer of the mucous membrane and of the uterine glands, as seen in certain rodents, help in their turn to prove to us that the laws governing the neo-formative process of the decidua, when it exists, and of the placenta in all cases, are identical in all mammals, whatever may be the form assumed by the placenta. I shall not enter into a minute analysis of this subject, reviewing separately my preceding observations at length, but shall limit myself solely to some brief considerations which seem to me sufficient for my present purpose.

In the case of diffused placenta, it is known that in its simplest form, described by me in the sow, and by Turner in *Orca gladiator*, every trace of the old epithelium of the uterine mucous membrane of the non-gravid uterus is entirely lost, and that a new epithelium of different form covers the small crypts of new formation in which are disposed the villi of the chorion. In the solipeds the shape of the new crypts or glandular follicles clothing all the inner surface of the gravid uterus is better defined and more complete than what was observed in the sow and the cetacea; evidently, however, the new formation takes its origin immediately below the old epithelium of the uterine mucous membrane of the non-gravid uterus, wherefore, in all cases, the destructive process of the former elements of the mucous membrane certainly affects in these animals the old epithelial layer throughout, and for the most part extends to the more superficial layer of the sub-mucous connective tissue. Over all the internal surface of the uterus thus laid bare takes place the neo-formative process, at first of cellular elements only, from which as pregnancy advances develop the crypts or follicles that constitute the maternal portion of the placenta. In animals having a diffused placenta a decidua vera does not exist, since it is all the neo-formed elements that, over the whole internal surface of the uterus, are transformed into the new glandular organ constituting the maternal portion of the placenta; or, if you please, the decidua vera, in the case of diffused placenta, is represented by the old epithelium of the mucous membrane which is detached; as in the rabbit the decidua vera is represented by all the elements of the mucous membrane, by the layer of submucous connective tissue, and the glands which are separated from the uterus after conception.

In those mammals having a diffused placenta, as in certain rodents with a single placenta, the neo-formative process is always preceded by a destructive metamorphosis of old elements of the mucous membrane of the non-gravid uterus, the only difference being that in diffused placenta the destructive changes are confined to the epithelial layer that covers the uterine mucous membrane, and the neo-formative process is extended and completes itself over the whole inner surface of the uterus; while, on the other hand, in cases of single placenta, the destructive process attacks extensively and more deeply the entire uterine mucous membrane; but the neo-formative process, uniform at first, is afterwards always limited and circumscribed to that point only of the uterine surface where the placenta is attached. The decidua vera is composed of the new elements that are arrested in their development when the first phase of the neo-formative process takes place, as we have seen with certainty in the rat.

The multiple placentæ which I have studied in the cow, in the sheep, and in the deer are developed only at certain fixed points of the inner surface of the uterus, namely, in the cotyledons of the non-gravid uterus, and it is at these places only that the destructive and superficial changes occur. The neo-formative process, by which the uterine cotyledon of the gravid uterus is produced, is also limited to these same points. In these cases, likewise, a decidua vera is not observed. It does not exist on the uterine surface unoccupied by the cotyledons, because in this extensive portion of the uterus the parts which constitute the mucous membrane remain intact; it is not found in the cotyledons, since at those places the active proliferation of cells from the walls of the neo-formed vessels constitutes the placental masses or uterine cotyledons, which have been regarded by many, and recently even by Kölliker, as simple tumefactions of the uterine mucous membrane. In the case of multiple placenta, also, the neo-formation is preceded by a metamorphosis of some epithelial elements limited to certain places only of the uterine mucous membrane of the non-gravid uterus. In short, there occur at circumscribed points of the internal surface of the uterus the same changes which we have seen taking place over the whole of that surface in animals having a diffused placenta.

Observations which I have made on the formation of the cotyledons in the cow seem admirably to confirm what I have just stated. It is generally thought that in the cow, during pregnancy, the neo-formative process takes place exclusively in the circumscribed spaces known under the name of cotyledons of the non-gravid uterus, but whoever examines the gravid vaccine uterus will not seldom find in the midst of the large, normal cotyledons small ones scattered here and there, from the size of a grain of millet to that of a bean, or larger. The true signification of these small cotyledons was rendered plain to me by a beautiful anomaly noticed by Dr. Rossi, municipal veterinary surgeon. In a gravid uterus of the cow, where the inner surface was covered with these small cotyledons, and in some parts so extensively as to appear diffused in the midst of the normal ones, the anatomical structure (size excepted), even with regard to the foetal portion, was the same in the small irregular and in the larger normal cotyledons.

Now these facts, in my opinion, show that in the uterus of the cow, at the time of conception, the destructive process of the epithelium is not always limited to the spaces occupied by the cotyledons, but extends to some points of the remaining uterine mucous membrane, and that in the instance mentioned it was exceptionally extended to a great part of the inner surface of the uterus; also, that in the special condition of the uterus at the moment of conception the occurrence of the epithelial denudation is sufficient to establish the neo-formative process, or the development of small cotyledons even on that portion of the uterus where they would not be developed normally. Accidentally in the uterus of the cow may be formed cotyledons of the gravid uterus outside of the places preëstablished by nature (cotyledons of the non-gravid uterus), which I have elsewhere shown takes place normally in the uterus of *Cervus Porcinus*.

The observations mentioned, both the normal and anomalous, show that the establishing of the neoplasm on the part of the mother is all that is needed in order that from the opposite side the fœtal portion may be developed in the chorion; and the anomaly observed in the cow explains how in the normal state the fœtal cotyledons are formed in the chorion only, and precisely in the places corresponding to the maternal cotyledons.

In the carnivora, as in the quadrumana and in woman, the destructive process is equally superficial, and limited to the epithelial layer; but in the carnivora, as in other multiparous animals, it is confined to those segments of the uterine horns where the ova have been arrested, while in the quadrumana and in the human species it affects the epithelium of the whole cavity of the uterus.

In these cases the neo-formative process is determined by the extent of destructive changes; that is, it is limited to the segments of the uterine horns where the ova are found, or, on the other hand, extended to the whole inner surface of the uterus.

The neo-formative process, which is established wherever the uterine surface is denuded of epithelium, gives rise to the formation known as decidua, and whether this becomes vera, reflexa, or serotina depends on the place where the ovum is arrested and where the placenta will be formed.

In advanced pregnancy, or at its close, the decidua vera represents the first phase of the neo-formation, which was arrested in its development because the placenta was formed at another part of the uterus, since at the place where the placenta was formed there is proof that the first neo-formative process continues and completes the phases of its growth, constituting the serotina.

The activity of the first destructive process and of the subsequent neo-formative one that uniformly follows in the segments of the horns of the uterus in the multiparous females, or over the whole internal surface of the uterus in the quadrumana and the human species, may be estimated by the thickness of the decidua vera before this is detached from the internal wall of the uterus, because it represents the two facts which I have indicated above. We have observed these changes with greater exactness in the gravid uterus of certain rodents where it is the whole mucous membrane and also the thick glandular layer that are altered and detached in the first phase of the destructive process; but in the human species and in the rodents that we have examined where the epithelial layer alone is separated, the decidua vera is represented by the neo-formed cells and the elements of the mucous membrane that have been affected by the destructive process, or by the last alone, as we have seen in the rabbit.

Although, indeed, in woman only the epithelial layer is separated, yet it is the innermost layer of the muscular tissue of the uterus that remains uncovered, as we have seen to be the case in the rat, in the rabbit, and in the cavia; so that, notwithstanding the considerable differences we find with respect to the parts attacked by the destructive process, the change that takes place in the cavity of the uterus is identical in all cases: it is always the internal muscular tissue of the uterus that remains uncovered and denuded.

In the human species, as in the rodents examined, after the separation of the decidua vera from the inner surface of the uterus and during the period of pregnancy, the internal muscular surface of the uterus is clothed with a uterine mucous membrane identical with that with which it was provided before conception. In woman, the reproduction of the simple epithelial layer, which by itself represents all the mucous membrane of the uterus, is all that is needed in order that the internal surface of the uterus should be restored to the primary normal state that preceded the pregnancy. In the rodents the regenerative process is much more extensive; the epithelial layer is indeed the first to be formed in these animals; but to return to the primitive state there must be a development of the utricular glands, and a renewal of all the anatomical elements that are found in the sub-mucous connective layer which normally is superimposed on the uterine muscular tissue.

These summary considerations sufficiently prove, as I think, that the changes which are established in the uterine mucous membrane immediately after conception, while varying in degree, in extent, in importance, do not vary in their intimate nature, which consists in a primordial destructive process, more or less marked, of the uterine mucous membrane, which is indispensable for the beginning and completion of the neo-formative process, from which, in all cases, results the development of the maternal portion of the placenta. The unity thus maintained amid so great diversity of the conditions which, so to speak, govern the formative process of the placenta in the different mammals helps now in our inquiry on the unity of the anatomical type of the placenta in the mammalia and in the human species, in connection with the physiological unity that controls the nutrition of the fœtus in all vertebrates.

II.

It is a well-known fact that during the period of embryonic life, the vertebrate animals, to whatever class they belong, in order to complete the marvelous phases of their development, require a special nutriment, which is always furnished them by the mother, and this whether the embryonic life is perfected apart from the body of its mother, in the interior of an egg, or, on the contrary, within, by a direct relation with the uterus of the mother.

Indispensable, therefore, in all cases, for the nutrition and the development of the fœtus in the vertebrate animals, are both the maternal aliment and the means by which it may be appropriated and converted into its own substance by the embryo; and however many and considerable may be the differences met with in these two fundamental factors, the unity running through them readily appears under two general forms, represented by the yolk of the egg in the oviparous animals and by the placenta in vertebrate mammals. The only difference in these two facts, at first sight so dissimilar, consists in this: that the nutritive material for the embryo of the mammalia is furnished by the mother by means of a special organ, the placenta, which elaborates it by degrees, as the embryo is developed; whilst in the oviparous vertebrates, the maternal aliment is stored up in a mass by the mother, and emitted with the egg in the quantity needed by the embryo to complete its growth. The means available to the embryo for appropriating the maternal aliment are always, in every case, furnished by its vascular appendages, which are destined to absorb and convey to the fœtus the materials needed for development.

Considering the question from this general point of view, the two fundamental parts which constitute the placenta of the mammifera, the maternal and the fœtal, are also found in the oviparous vertebrates, and therefore, strictly speaking, a perfect placenta in its two basic divisions cannot be denied to these vertebrates, because the growth of the embryo, both of the mammiferous and of the oviparous animals, is governed by one and the same law, and the great diversity of the methods adopted by nature in the two is subordinate to the unity of the mode in which she accomplishes her purpose, namely, the nutrition of the embryo from the materials which are furnished by the mother. But, more than this, even anatomically, the connecting link between the oviparous vertebrates and the mammalia is not wanting. Huxley[1] found that in certain fishes, especially in the viviparous selachii of the genera Mustelus and Carcharias, there exists a sort of rudimental placenta, formed by introflexions or numerous folds of the uterine mucous membrane, corresponding with as many similar ones in the walls of the umbilical sac. I am not aware that this important observation has been repeated and carefully investigated by others, and shall therefore give the results of my own studies; but I ought

[1] Manuale dell' Anatomia degli Animali vertebrati, page 125. Florence, 1874.

to mention that analogous observations were long ago made, and have been recently confirmed by able and minute researches, which are too important to be omitted.

As early as 1787 Cavolini[1] stated that he had seen in a gravid torpedo (*Raja torpedo*, L.) the embryo almost developed in the uterus, with the yolk still attached by means of the umbilical cord; but this yolk was fastened to the interior of the uterus by means of an infinite number of red papillæ, existing upon the uterine wall and clinging to the body of the yolk. I endeavored to repeat this observation, which would have been of special importance in explaining this part of my work, because the papillæ or uterine villi indicated by Cavolini would represent in certain viviparous fishes the elementary typical form, which I shall show exists in the maternal portion of the placenta of the mammalia. But if the statements of Cavolini are verified upon examination only of the gravid uterus of a torpedo, there arises some doubt as to their importance when we compare the gravid with the non-gravid uterus of this fish, since the mucous membrane of the non-gravid uterus is likewise covered with an infinite number of villi, which, in volume, form, and the anatomical elements composing them, are identical with those observed in the gravid uterus. One sole difference is noted, which, however, explains the observation made by Cavolini, and this is, that while the mucous membrane of the non-gravid uterus of the torpedo is uniformly covered with a villous coat, in the gravid uterus, on the other hand, this is found accumulated at certain

[1] Memoria sulla Generazione dei Pesci e dei Granchi. Naples, 1787.

points, and restricted to these, considerable portions of the mucous membrane being destitute of villi, or showing but few, and those far apart from each other. Any doubts, however, that may arise when limiting the investigation to the uterus of the torpedo are completely dispelled when the observations are extended to other fishes, as was done by Bruck in his important work,[1] in which he not only quoted the researches made by J. Müller and Leydig on the disposition of the folds of the mucous membrane and the presence of numerous and large villi in the gravid uterus of different species of selachii, but by accurate examinations of his own proved that these numerous villi are developed only after conception has taken place.

He considers[2] the uterine mucous membrane the most important part for study, because in the viviparous species it is the seat of interesting modifications at the period of gestation. In the non-gravid uterus the mucous membrane is of a pale red color, very smooth, appearing more velvety near the posterior extremity. Before the ovum arrives in the uterine cavity, the epithelium that covers the mucous membrane proliferates and the velvet-like appearance is more evident, because the wall of the cavity acquires a plainly villous aspect. In *Pteroplatea altavela*, the villi are so numerous that the mucous membrane is entirely covered; they are from one to two centimeters long, from one and a half to two centimeters wide, and are so close together and so inter-

[1] Études sur l'Appareil de la Génération chez les Sélaciens. Strasburg, 1860.
[2] Idem, p. 58, etc.

twined in all directions as to form an inextricable mass, soft and pulpy. When the uterus incloses one or two fœti the villi are so long and so numerous that the young seem as if hidden in a vascular nest. When the villi are fewer in number they present an arrangement more or less regular; according to Leydig they are disposed in longitudinal rows, very regular, in *Scymnus lichia* and in *Acanthus vulgaris;* but in order to observe these dispositions, it is necessary, Bruck tells us, " to study the uterine mucous membrane at the beginning of gestation, since at a later period the villi are so long and so involved that all regularity seems lost."

These villi are composed of the same anatomical elements that form a villus in any mucous membrane whatever, and according to the author are developed to form a vascular envelope for the young, in which they shall find the materials necessary for their growth and development.

Through the courtesy of my esteemed friend, Professor Salvatore Trinchese, I was recently enabled to study the uterine horn or chamber of a *Mustelus lævis* at the close of gestation, in which I found six perfectly formed young. After several careful examinations there was clearly shown the singular fact, which I have not seen mentioned, that every fœtus with its envelopes was included and perfectly inclosed between two large folds of the uterine mucous membrane, which had established an epithelial adhesion by simple contact, very close at the edges and throughout their whole length. The external membrane of the ovum, completely anhistous, and in direct relation with the internal surface of the two folds forming the completely closed external sac, was covered with a somewhat dense layer of mucus. The long umbilical cord terminated in an ample umbilical vesicle, and this at its extremity adhered strongly to an elevation of the uterine mucous membrane, oval in shape, and about a centimeter in diameter. The uterine mucous membrane as well as the umbilical vesicle was at this point very richly furnished with vessels. I did not observe upon the surface of the umbilical vesicle the vascular appendages mentioned by Huxley.[1] This species of rudimental placenta is also found inclosed in the sac formed by the folds of the mucous membrane, and precisely at the lower angle made by them, as they project and comprise the whole ovum in their interior.

Huxley had already taught that in the aforesaid species of mustelus, and in others of carcharias, there was observed this sort of rudimental placenta, formed by the doubling of the walls of the umbilical sac and of the uterine mucous membrane; and this kind of introflexion, as I have been able to verify, really exists, is very close and complicated, and the superficies of contact between the epithelium of the umbilical vesicle and that of the uterine mucous membrane is therefore by far more extensive than had been supposed. These observations made on certain viviparous fishes are, in my opinion, of no little interest, because they represent in the most simple and rudimental way the same forms which are found in the placenta of the mammalia.

The contact of the uterine mucous membrane with the foetal envelopes which I have noticed in *Mustelus*

[1] Op. cit., p. 32.

lævis may perhaps have its counterpart in the placenta of the marsupialia; but assuredly the simplest forms of diffused placenta of certain mammals, with a much higher and more complex structure, find their counterpart in the elementary form indicated in *Mustelus lævis*, while, according to the observations of Bruck, in other viviparous fishes we have, as we shall see, the primordial type of the maternal portion of the more perfect and complicated placenta of mammals, and even of the human species.

It is sufficient now to notice that a very important link between the oviparous and the mammiferous animals is established, and that, in the most elementary or primordial forms of placenta which are met with in certain viviparous fishes, there are found clearly distinct the two fundamental parts of the placenta of the mammals, namely, the maternal and the fœtal. Later we shall ascertain that even in these elementary forms the placenta preserves the anatomical type of the organ, which is observed in its highest development in different mammals and in the human species.

The inaccurate and often erroneous opinions hitherto held upon the structure of the placenta in mammals have not been limited to the field of anatomical or physiological science, but have embraced a much wider sphere, since able men like Owen and Huxley, and very lately Kölliker, have marked out fundamental distinctions among the mammiferous vertebrates, drawn from the presence or absence of the placenta, or from the occurrence of certain facts in the act of parturition, joined with the general forms which the placenta assumes in the different orders of mammalia.

All know that so illustrious a scientist as Owen taught this distinction among the mammals: that there were those without a placenta (mammalia implacentalia), and those possessing a placenta, which he called mammalia placentalia.

Among the implacentalia Owen placed the marsupialia, and direct observations on the monotremia being wanting he inferred that they too should be assigned to the implacentalia.

Recently Kölliker[1] has proposed to substitute the name of mammalia achoria for Owen's implacentalia, and that of mammalia choriata instead of placentalia. In his opinion, many mammals which, according to Owen, are placentate, and by many are improperly described as having diffused placenta, have none, and he explains his idea by stating[2] that " in those animals incorrectly classified as having a diffused placenta, the chorion is covered with small villi, which penetrate into simple cavities of the uterine mucous membrane, from which they are easily removed; hence formations that may be regarded as analogous to a placenta are entirely wanting;" and, treating of the placenta of the ruminants, he remarks that in these animals[3] " the villi of the chorion have many ramifications which dip down into spacious hollows of the tumefied uterine mucous membrane, and, moreover, the fœtal and maternal parts stand so related as to constitute numerous formations analogous to placentæ."

According to Kölliker, then, not only would the marsupialia be implacentalia, and perhaps the mono-

[1] Entwickelungsgeschichte des Menschen und der höheren Thiere, page 352. Leipzig, 1876.
[2] Ibidem, page 354.
[3] Ibidem.

tremia, as was taught by Owen, but also the cetacea; of the pachyderms, the elephant, the hyrax, the hog, and the solipeds; among the ruminants, the comelidæ, as well as the genus manis among the edentata, — all of them animals in which the two fundamental parts of the placenta have been of late, by Turner and myself, in their most important particulars, minutely described and illustrated by numerous figures. I do not know if, when he says that in ruminants the cotyledons " constitute formations analogous to placentæ," Kölliker believes that these animals are not really placentate, a point with regard to which doubt is no longer permissible. At all events, the subject is important enough for us to inquire how a man so distinguished in science was led into so grave errors.

In order to ascertain the structure of the placenta, this illustrious histologist made the study of the human placenta the foundation of his researches, and owing to the great difficulties which are met with in its investigation with the ordinary anatomical methods he fell, or rather he was drawn, into the error of asserting[1] "that it was merely for the convenience of description that one designated the human placenta as maternal and fœtal." This imperfect knowledge of the two fundamental parts of the placenta led him into the further error of regarding as simple depressions or tumefactions of the uterine mucous membrane those parts which have been shown by Turner and myself to be glandular follicles of new formation, constituting the secretory organ or maternal portion of the placenta, clearly distinct from the fœtal and absorbent portion, composed of the villi of the chorion. Turner,

[1] Op. cit., p. 331.

with great clearness, has arrived at this conclusion when he states[1] that no precise idea or exact conception can be formed of the intimate structure of the placenta except by considering the two surfaces in their relation to each other, to wit, the maternal or secretory surface, and the fœtal or absorbent.

As a necessary consequence of error, many animals would be considered by Kölliker as without placenta, when their possession of it has been proved beyond doubt, and that it is formed by the two fundamental parts which are indispensable in all vertebrates for their nutrition and for the completion of their period of embryonic life.

I claim indispensable, although the illustrious Owen, from personal observations, maintained that in the marsupialia the placenta is wanting, and it was precisely upon these observations of fact that he founded the great distinction of mammifera implacentalia and placentalia.

Eliminating the erroneous amplification which Kölliker sought to introduce into mammalia implacentalia, there remains the distinction founded by Owen, which indeed rests upon a single observation that he made upon the uterus of a pregnant female of a gigantic kangaroo, or *Macropus major*. This celebrated anatomist declared that the fœtus was wrapped in an external membrane, rather delicate and without vessels, which had the appearance of being, and probably was, a serous membrane; that below this was the amnion; the umbilical vesicle was furnished with large omphalo-mesenteric vessels com-

[1] Lectures on the Comparative Anatomy of the Placenta, page 114. Edinburgh, 1876.

municating with the vessels of the amnion, but there was no trace of the allantois or its vessels. It ought, however, to be added that Owen himself, when he had occasion to examine a young kangaroo that had descended a little way into the marsupion or maternal pouch, observed in it the urachus, which extended from the vesicle as far as the umbilicus, and in this only two umbilical arteries in the interior, without the corresponding veins. From these observations, he admitted that in the kangaroo, towards the end of intra-uterine life, may be formed a small allantois, but that there is no relation established by it between mother and fœtus.

However great the respect I hold for so eminent a scientist, yet it cannot be denied that the very facts that he has advanced give rise to grave doubts as to the entire absence of the placenta in these animals — doubts necessarily resulting from his own observation of the presence of an umbilical cord, although it was formed only by two umbilical arteries.

If the illustrious Turner had not confirmed my observations on the maternal portion of the placenta in the sow and the mare, and had not extended them to the cetacea in his description of the placenta of *Orca gladiator*, I might feel some hesitation in mentioning that, at the time when Owen examined the uterus of the kangaroo, my own studies and those of Turner were unknown; and also the important changes which the mucous membrane in these animals undergoes at the beginning of gestation, to give rise to the neo-formative process of the maternal portion of the placenta had not been described. Therefore, the observations upon the embryo of the marsupialia, which led Owen himself to infer that even in these animals, at the close of intra-uterine life, an allantois might be formed, gave grounds for the supposition, from facts afterwards learned, that although in the marsupialia also (less completely and later, indeed) there may take place upon the mucous membrane the formation of the maternal portion of the placenta, and villi may be formed upon that serous envelope which was without them at the period of gestation, when Owen made his observation, yet this neo-formative period of the two parts of the placenta may be more delayed, and consequently of shorter duration, in the marsupialia than it is in other mammals having diffused placenta; and the presence of the two umbilical arteries noticed by Owen in a young kangaroo that had descended into the maternal pouch considerably strengthens such an inference.

Perhaps the form of diffused placenta found in these animals may be more elementary and simple than even that observed in the hog, on account of the peculiarities attending birth in the former, and therefore not easily noticed in a first or single examination. But the facts, so far as known, rather confirm than weaken the deduction that also in the marsupialia there exists a placenta composed of its two fundamental parts, namely, the maternal and the fœtal; and until our knowledge on this point shall be clear and certain, we cannot admit that it has been proved that the ovum of the marsupialia does not receive for its development nutrition from a placenta, which, as I have remarked, cannot, strictly speaking, be denied even to oviparous vertebrates, and of the existence of which in certain viviparous fishes we have positive proof.

But even allowing as indisputable the demonstration that every kind of placenta is wanting in the marsupialia, they yet could not be styled implacentate, because there would still remain the fundamental fact characteristic of the placental organ, namely, the contact of two surfaces, — the fœtal or absorbent, represented by a chorion without villi, and the maternal or secretory, represented by the uterine mucous membrane, which I have proved exists as a rudimentary form of the placenta in the *Mustelus lævis*.

From my earliest observations on the different forms of placenta in certain mammals, I stated the reasons for my conviction that the fluid secreted by the utricular glands serves for the nutrition of the fœtus at the beginning of embryonic life, in the period, that is, which precedes the development and the vascularization of the villi of the chorion, and the formation of the crypts or glandular follicles in the case of diffused or multiple placentæ, or of the maternal portion of the placenta where that organ is single under the zonarial or the discoidal form. Turner[1] was induced to believe, from the consideration of the vascularization of the chorion even in the extended zones or areas of it which in certain animals are destitute of villi, while the utricular glands, which stand opposite in the uterine wall, are increased in volume, that the vessels of the chorion in these zones were the absorbents of the materials furnished by the utricular glands during the whole period of gestation. He was not,

[1] Op. cit., p. 117.

however, able to decide positively as to the special office of the glandular secretion in the nutrition of the fœtus.

In certain animals having a diffused placenta, as in the solipeds, and in others with a multiple placenta, like the sheep, the inference of Turner can in no wise be disputed. In the solipeds, especially, it has been shown that the lacteal fluid that bathes all the uterine surface of the chorion must result from a mingling of the secretion of the utricular glands with that separated by the glandular follicles or crypts of new formation which constitute the maternal portion of the placenta. But in the gravid uterus of the sow, where the placenta is diffused, in the mare and in the cow, where it is multiple, in the sheep, where it is pluri-cotyledonal, I observed and described some anatomical facts which seemed to certify that in the latest period of gestation, against the aperture of the utricular glands, were constantly formed certain obstacles which prevented the free flow between the uterine surface and the chorion of the fluid secreted by the utricular glands.[1] From these observations it appeared to me reasonable to believe that the importance of the secretion of the utricular glands for the nutrition of the fœtus might, in some animals, be limited to the earliest periods of their embryonic life, while in others it might continue during the whole term of fœtal life.

An analogous contradiction is noticed in different animals having the placenta single, and which by

[1] Previous to my observations, Burckhard, in his memoir entitled De Uteri Vaccini Fabbrica, had already noticed the formation of this obstacle in the uterus of the cow.

recent observations, even, has been brought out in a somewhat important way.

In the greater number of cases where the placenta is single and in the human species itself, by the ample openings which the utricular glands maintain into the decidua vera, the inference drawn by Turner cannot be disputed. But the examinations which I have made upon the formation of the decidua and the placenta in certain rodents, in which all the old uterine mucous membrane, including the sub-mucous vasculo-glandular layer, is very quickly and altogether destroyed, in order to give place to the neo-formation of the decidua and the placenta, do not allow the belief that in these animals the utricular glands by means of their secretion have any importance in the nutrition of the fœtus. This is especially true in these animals, since the whole uterine surface not occupied by the placenta, and which is below the decidua vera of new formation, is clothed with a simple epithelial layer, which by itself represents, during gestation, all the elements of the uterine mucous membrane, the glands, and the underlying vessels which existed on the internal surface of the uterus previous to conception. These observations prove beyond all doubt, in my opinion, that the fluid secreted by the utricular glands is not indispensable for the growth of the embryo in the mammifera, and much less is it indispensable to it through the whole period of intra-uterine life.

Notwithstanding this conclusion, the observations made and the inferences drawn by Turner and myself clearly show that if in some animals the fluid secreted by the utricular glands does not serve for the nutrition of the fœtus even in the beginning of utero-gestation, yet in others it may in some manner aid in nourishing the fœtus during the whole period of intra-uterine life, while in most animals it may be the sole aliment that the ovum, separated from the ovary, finds in the uterus of the mother, and which of itself suffices, for a longer or shorter time, for the nutrition and development of the embryo.

This, as is evident, does not imply a very wide dissension between the eminent Turner and myself, but merely a different appreciation with regard to the time in which, in the greater number of instances, the secretion of the glands is necessary for the nutrition of the fœtus. The importance of their secretion in the early period of pregnancy, or in some cases during the whole time of gestation, is not affected, and can in no way be absolutely excluded by the new facts which I have observed in the rodents.

Now, as all are aware, the embryonic phases which in the mammifera are accomplished in the maternal uterus, are in the marsupialia divided into two distinct periods : that of intra-uterine gestation, which is very short, and the rather long period of marsupial gestation, which comprehends the latest phases of the embryonic and the first period of the extra-uterine fœtal life of the other mammals. Therefore, in the marsupialia, during the short period of their intra-uterine life, the fluids secreted by the utricular glands may suffice for nutrition and development ; and from observations made on other mammiferous animals this may be held as probable, until positive proof to the contrary is given, which has not yet been done. At all events, the want of villi in the chorion, as well as the absence of the crypts or glandular follicles during gestation, is not sufficient ground for pronouncing the marsupialia without placenta. There may be a difference of form, but not an absence of the placenta, its two fundamental parts, the absorbent or foetal and the secretory or maternal, being distinctly represented. In short, in the marsupialia the secretion of the utricular glands is observed as exceptional and fundamental, which in other mammals generally is transitory; and even though it remains of use through the whole term of pregnancy, it is still associated with a glandular organ of new formation which is invariably present. In the marsupialia, then, we have in a higher degree the repetition of the fact observed by me in the *Mustelus lævis*.

We cannot therefore now positively assert that this new organ in its most elementary form is wanting in the marsupialia, or that it is represented only by an increase in volume and secretion of the utricular glands, which has been shown to take place in the gravid uterus of these animals; but it may be that in both these ways there is found in the marsupialia the simplest form of placenta among mammals, joined with the simplest but relatively higher form of diffused placenta seen and described by Turner in the cetacea, and by myself in the sow.

As early as 1829 the illustrious Cuvier had noticed that the marsupialia constituted a separate class, parallel with that of the mammifera, the insectivora, the rodentia, the carnivora, the ruminantia, and the quadrumana all meeting in it. The opinion of Cuvier, confirmed afterwards by the discoveries of Owen and others in fossil marsupialia, would acquire a fine demonstration should it be proved that the foeti of the marsupialia have a large umbilical vesicle, from which vessels extend to the chorion, which is folded.[1] Who does not perceive the importance of such an observation, when placed beside those recorded above, on the most elementary forms of placenta found in certain fishes, in which the fœtal nutritive vessels originate from the umbilical vesicle, and not from the allantois? The marsupialia would constitute a distinct class, as Cuvier taught, which by its mode of placental formation would be the connecting link between the viviparous fishes and the mammalia.

But it is not in the marsupialia, where our knowledge of the placenta is so scanty and uncertain, nor in the monotremia, about which information is altogether wanting, that we are to look for the unity of the anatomical type of the placenta; such inquiry, to be useful, must be directed to those mammals in which the existence of the placenta cannot be doubted, whether diffused, multiple, or single, with either a zonarial or a discoidal form.

To these three principal forms of placenta correspond three fundamental facts, belonging to its anatomical structure, which may be divided into two general classes. In the first may be placed all the forms of placenta in which the glandular structure of the maternal portion is plainly evident; in the second those in which the glandular structure is obscure.

The first division comprises diffused and multiple placentæ: diffused, when the glandular organ of new formation takes the form of crypts, or of simple glandular follicles; multiple, when it assumes that of a

[1] Huxley: Manuale dell' Anatomia degli Animali vertebrati, page 337. Florence, 1874.

compound glandular organ. In the second division, the glandular structure of the maternal portion of the placenta is concealed, as in all cases of single placenta, from the fact that the vascular or absorbent portion of the placenta is in direct connection with the cells that constitute its maternal, glandular, or secretory portion. The close relation in these cases of the two parts of the placenta has been the only obstacle met with in forming an exact idea of its intimate structure.

The observations which I have made upon the intimate structure of the placenta of the *Cavia cobaya* have proved that even in cases of single placenta there may be found united in one animal the visible glandular form and the hidden in the maternal portion of the placenta; and this union of two forms so different in one placenta has seemed to me to possess a certain importance, as showing in close relation what we have believed most dissimilar.

Professor Turner, in his important work,[1] came to the conclusion, elsewhere pointed out by me, that to form an exact idea of the anatomical structure of the placenta two surfaces are to be considered, — the fœtal or absorbent, and the maternal or secretory (or, as I have named it, glandular). This fundamental fact can no longer be regarded in any degree doubtful, but, formulated in so general a manner, this idea of Turner's cannot be applied with certainty to explain the manifold differences met with in studying the anatomical structure of the placenta in animals and in the human species.

---

[1] Lectures on the Comparative Anatomy of the Placenta, page 114. Edinburgh, 1876.

Comparing the observations which I have been able to institute with those very carefully made by Turner on many other animals, I think I have arrived at an ultimate conclusion, and have been able, in the midst of the manifold and striking differences presented by the diverse forms of placenta in the mammals, man included, to fix upon a typical and fundamental anatomical form of placenta, common to all. If I have demonstrated the truth in this new deduction, which formulates anatomically the physiological notion accepted and learnedly illustrated by Turner, a somewhat singular consideration will be presented. Hitherto, the ablest anatomists, as we know, have been confused by the diverse structural appearances seen in the placenta of the various classes of mammals, and the differences seemed so wide and so important as to mislead the most learned scientists.[1]

Keeping in mind this fundamental anatomical unity, on the contrary, it will no longer be the remarkable differences that command attention, but we shall be impressed and surprised that with the few and simple means adopted by nature there should be so great a variety and multiplicity of forms as to conceal until now the unity of anatomical structure of the placental organ.

Leaving for the present the question relative to the formation of the placenta in the marsupialia and the monotremia, for the solution of which data and exact observations are wanting, although, as I have previously remarked, the idea of contact of the secretory and the absorbent surfaces may furnish an accurate

---

[1] See Kölliker: Entwickelungsgeschichte des Menschen und der höheren Thiere, page 352. Leipzig, 1876.

conception of its structure in those animals, we have for consideration all the clearly determined forms of placenta known to us, from the simplest, which is the diffused placenta in the sow, to the most complex, the single and discoidal in the quadrumana and in the human species.

The typical anatomical form which I have noticed in the two fundamental parts of the placenta, whatever may be the outer appearance, is identical in all cases: it is that of a villus of new formation, both in the maternal and in the fœtal part, composed of an internal vascular loop, surrounded by protoplasm, which is furnished with an outer epithelium. The office only of the two villi is different, the fœtal one serving for absorption and the maternal one for secretion.

The two villi always come in contact with each other. This contact may be more or less intimate, but the walls of the two vascular loops never, in any instance, touch; and this is the most important fact, because it invalidates the ideas hitherto held as to the mode of nutrition of the fœtus in the uterus. In the case of diffused and of pluri-cotyledonous or multiple placentæ, the secretory epithelium that covers the vascular loop of the maternal villus is found separated, and only in simple contact with the absorbent epithelium of the fœtal villus, and this is why in placentæ of these kinds the appearance of a glandular organ plainly results. According to the different ways in which the secretory villi are united or distributed upon the internal surface of the uterus are constituted the crypts or the glandular follicles, both simple and complex, which have been described by Turner and myself, and which Kölliker[1] even now regards as multilocular depressions or tumefactions of the uterine mucous membrane, into which the fœtal villi enter.

In cases of single placenta, whether of zonarial or discoidal form, the glandular character of the organ, as a whole, remains concealed, as I have stated, because the relation between the two species of villi are very closely maintained. Although the epithelium of the maternal secretory villus is always present, that of the absorbent villus is wanting, on account of the intimate relation between the two villi; the vascular walls of the fœtal loop can of themselves better fulfill their office of absorption by tenaciously adhering to the secretory epithelium which clothes the maternal villus. Accepting the idea of the typical and single form of the two fundamental parts constituting the placenta, it is readily perceived that all the manifold and apparently essential differences in the very simple structure of the two villi that are met with in examining the diverse forms of placenta in the mammalia depend upon three factors only. These factors may be thus specified:—

(1.) The manner, as has been stated, in which there is established either an intimate relation, or else merely a simple contact, between the two villi constituting the two fundamental parts of the placenta.

(2.) The absence, as I have shown, of the epithelium in the vascular loop of the fœtal or absorbent villus, because of the intimate relation which is established, in cases of single placenta, between the vascular loop of the absorbent or fœtal villus and the secretory epi-

[1] Op cit., p. 363, 364.

thelium, which is never wanting, in the vascular loop of the maternal villus.

(3.) Finally, on the form of the vascular loop in the interior of the maternal villus, which may maintain the usual regular diameter throughout its whole course, or may, on the other hand, offer dilatations or ectasiæ, which in the placentæ of the quadrumana and the human species are actually enormous.

A minute comparison of the condition of the three above-mentioned factors in the different forms of placentæ hitherto known and described would require too much labor, and too much repetition of what I have heretofore had occasion to mention. But, as I am anxious to prove as well as I can the truth and the simplicity of what has been asserted, I have placed together in the two divisions of Plate V. the diagrams of the different forms of placenta observed. This seems to me the most suitable way of explaining and demonstrating the unity of the anatomical type in all the various forms of placenta, and of showing that all the striking differences met with depend solely on the three simple factors which I have just pointed out. In the upper division, A, I have represented the plans of the diffused and the multi-cotyledonous placentæ; and in the lower, B, the single zonarial or discoidal, including the human placenta. The drawings in both exhibit a complete perpendicular section of the uterus, and of the two fundamental parts constituting the placenta; the letters in small type show the same parts in the two divisions; the numerals serve to indicate the plan of that form of placenta intended to be represented.

Thus, at *a* are represented the walls of the uterus,

whatever may be the form of placenta drawn in diagram above. I have omitted any representation of the utricular glands, because they vary in condition, and, as we have seen, are either completely destroyed before the placenta is developed, or all remain, acquiring a greater volume in cases of diffused and multiple placentæ, or they remain and augment in volume over all the surface of the uterus except the place where the placenta is formed. In this place, when it is single, they are altered and changed in shape, forming in woman that species of trabecula with large interstices which is met with between the inner wall of the uterus and the uterine surface of the placenta.

At b is indicated the layer of cells forming in all cases the basis of the uterine surface of the placenta, or the old cells of the decidua serotina in the human species.

At c is shown the innermost layer of the rich vascular net-work formed by the uterine vessels, so greatly increased in volume in the gravid uterus.

At d are indicated the placental vessels, properly so called; their relation will be shown when we come to explain the plan of the single placenta.

At e is represented that layer of cells which everywhere invests the placental vessels in all forms of the placenta. These are the parts which are situated upon the internal surface of the gravid uterus, and are in relation with the fœtal portion of the placenta.

The chorion is indicated at f; at g are the vessels that traverse the chorion in order to form the umbilical cord; and, lastly, at h are shown the vessels of the fœtal villi of the chorion itself.

Coming now to the different forms of placenta, the

numeral 1, of division A, represents the fundamental typical form of the fœtal portion of the placenta, consisting of a vascular loop communicating with the vessels of the chorion, and covered with its absorbent epithelium. At 2 is seen the typical form of the maternal portion, similar to No. 1 except that the vascular loop is in direct relation with the uterine vessels, and the external epithelium that clothes it is in direct continuity with the cells of the serotina. We thus notice that, while preserving the typical form in the two villi indicated, there may be obtained, by means of the three simple factors mentioned, the most varied forms of placenta, — from the simple and diffused, which has been observed in the sow, to the most compound and complicated, as seen in the quadrumana and in the human species. Nos. 3–4 represent the plan of the simplest form of diffused placenta, minutely described elsewhere in the sow; 3, the fœtal portion of the placenta in this animal, which does not differ from the first division of the diagram; 4, an exact plan of the maternal portion of the placenta in the sow, which is formed by the villi, regularly arranged, constituting numerous folds of the uterine mucous membrane, in the depressions of which the villi of the chorion are inclosed. Carrying the loop or fœtal villus of No. 3 between the two maternal villi, No. 4, we have the precise plan of a vertical section of the placenta of the sow, and of the cetacea, as was shown by Turner in the *Orca gladiator*. At 5–6 we have the plan of the diffused placenta of the mare; 5 indicates the fœtal portion. The absorbent villi do not deviate from the typical form in No. 1, except that instead of being isolated they are numerous, and joined together in a sort of tuft. In the maternal portion, No. 6, the secretory villi are also united, so as to constitute calyces or follicles, for receiving the loops of the tuft made by the villi of the absorbent or fœtal portion. From these calyces, formed by the secretory or maternal portion, and diffused over the whole internal surface of the uterus, there results a sort of layer, composed of so many crypts or glandular follicles of new formation, each of which receives within it a villus of the fœtal portion. Nos. 7–8 represent the plan of a multi-cotyledonal or multiple placenta, particularly that of the cow, — which is more complicated in form than that of the mare, but which yet presents the identical structure in both the fœtal and the maternal portions. The glandular follicles, No. 8, are not simple, but compound, and the fœtal villi have an arborescent form instead of a simple tuft, as in the mare. The peculiarities of structure which I found and described in the placentæ of the sheep, and of *Cervus axis* and *porcinus*, still remain plainly included in the diagram, although in these animals considerable differences might be observed on comparison with the placenta of the cow.

The fundamental typical form of the two villi seems to me clearly confirmed in the placentæ having a diffused form, and in those which are multiple or multi-cotyledonal. It is only the ideal unity which is wanting and must necessarily be wanting, since this unity results from analyzing into their elements the facts that determine the forms of placenta examined.

In all these forms the relation of proximity is evident, and of simple contact of the epithelium of the absorbent villus with the fluid furnished by the maternal villus, which assumes the form of a secretory organ. Although the typical form of the fœtal or absorbent villi is maintained distinct and clear in all cases, the form of the secretory villus is not on that account lost in the modifications assumed by the secretory or maternal portion of the placenta in the different species of animals which we have hitherto examined. In fact, in the sow and the cetacea (diagram 3-4) we have only a repetition, or an increase in number, of the typical villi in the fœtal as well as in the maternal portion. In a higher degree of development of this same form of placenta, that of the solipeds (Nos. 5-6), we find, besides an increase in number, an increase also of volume, in the two sets of typical villi, and the union of many maternal villi no longer gives rise to the appearance of small crypts, as in the former case, but rather to that of simple glandular follicles.

In the ruminants (Nos. 7-8), besides number and volume, there is a subdivision and ramification of the villi in place of the form hitherto under consideration, and therefore the form in the maternal portion is that of a compound glandular follicle. In all cases, however, the typical or primordial form of the two parts constituting the placenta is easily discernible. Apparently of far more importance are the differences in the examples of single placenta which I have arranged in the second division, B.

In all these placentæ the fundamental characteristic is that the epithelium of the fœtal villus disappears, in order that the walls of the vascular loop may come in direct contact with the secreting epithelium of the maternal villus, which is never wanting. This fact is represented in its simplest form in diagram Nos. 9–10. In the figure, the vascular loop of the fœtal villus, No. 9, appears slightly apart from the epithelium of the maternal villus, for the sake of avoiding confusion.

In fact, however, many single placentæ in different mammals appear to be formed only of a complicated web of maternal and fœtal vessels, which are always separated from each other by a cellular layer belonging to the maternal vessels. Now, in order that the diagram should be exact, the wall of the vessel of the fœtal villus, No. 9, ought to touch and adhere to the epithelium of the maternal villus, No. 10. Fig. 2, of Plate IV., which represents a section of the vascular portions of the placenta of the *Cavia cobaya*, shows the exact copy of a transverse section of the plan of placenta, figured at Nos. 9–10, as it is actually observed also in the placenta of other animals. Nos. 11–12 give the plan of the placenta in the dog, of which I treated in the first part of this work. Accurate drawings of this may be seen in Plate I., Figs. 2 and 3. This form of placenta, common to different carnivora, and which Turner has accurately described in the fox, is very well suited for demonstrating that the secretory epithelium, or the cells of the placental decidua, are elaborated from the walls of the maternal vessels, because the maternal portion of the placenta is formed, as I have said, of a large-meshed net-work of vasculo-cellular tissue (Plate I., Fig. 2). The course of the vessels is more or less winding in the placenta of different animals, and the sinuosity varies, also, according to the period of gestation, being greater or more frequent when the placenta has perfected its development.

In the diagram I represent at No. 12 the maternal portions, and the aforesaid net-work is reduced to two simple vasculo-cellular columns (d, e). In these forms of placenta, the vessels of the fœtal villi ramify into a rich and minute vascular netting, which is distributed upon the cellular envelope that surrounds the vessels of the maternal villus. This relation between the two portions of the placenta, which is accurately shown in Plate I., Fig. 3, is very simply exhibited in diagram No. 11, h indicating the rich capillary net, with only three branches for each trunk of the fœtal vessel. On comparing this with the preceding division of the diagram it is plainly seen that there is only a change in the mode of distribution of the fœtal vessels, and that also the fundamental fact, namely, of the vessel of the fœtal villus, without epithelium in contact with the epithelium of the maternal villus, remains unaltered in these forms of placenta.

Moreover in all single placentæ, whether they have a zonarial or a discoidal form (from which results the concealment of the glandular character in the maternal portion), another fact, exceptional, inasmuch as it is observed only in the placentæ of the quadrumana and the human species, is the ectasia of the vascular loop of the maternal villus. This constitutes one of the few modifications to be seen in the typical villi, and which has been the cause of the greatest and most serious difficulties hitherto met with in ascertaining the intimate structure of the human placenta.

In the fox Turner observed and described some

dilatations in the maternal vessels,[1] and with Eschricht looked upon them as a first intimation of the large lacunæ which are seen in the human placenta, and which I have also represented in that of the quadrumana. When describing the placenta of the *Cavia cobaya* I exhibited a beautiful net-work of ectasic capillary vessels, underlying the cotyledonous portion of the placenta in that animal. (Plate IV., Fig. 1, *n*, and Fig. 3, *a*.) That the lacunæ in the human placenta and the quadrumana were only enormous dilatations of the placental vessels I had already pointed out in my previous works, but the whole truth had not presented itself to me, because I had not been able to obtain positive knowledge that the cells of the decidua were elaborated from the walls of the maternal vessels. Comparative anatomy has now very clearly demonstrated this.

As we know, the illustrious Waldeyer referred the cells of the serotina to the group called by him plasmatic cells, having their origin from the connective cells, and being in intimate relations with the vessels. My own observations lead to the belief that the cells of the decidua are rather a perivascular cellular tissue.

The exact knowledge of this anatomical truth has a much greater importance than might at first appear. This I will now show by means of a diagram altogether imaginary, but which will presently find direct application when we inquire into the structure of the human placenta. This ideal figure is shown at Nos. 13–14, which is the same as the preceding diagram of the placenta of the dog, Nos. 11–12, with

[1] Op. cit., p. 85, and Pl. I., Fig. 1.

the addition of the imagined ectasis of the vessel of the maternal villus, which is represented in an incipient state on the left of the figure, — whilst on the right, the dilatation has reached its last stage of development. At the left the incipient and irregular ectasis of the maternal vessel has not yet carried the secretory epithelium with which it is clothed (e) into contact with the walls of the vessel of the foetal villi (h). At the right, on the contrary, this has taken place, as we plainly see it must, from the continued dilatation of the maternal vessel.

Now simply looking at this right side of the figure three things appear as truths, which are, in reality, only deceptive appearances. The first is, that instead of the primitive vessel there is formed a lacuna of blood (d), while it is only a real ectasis. The second deception is that the foetal villi (e) are furnished with a proper epithelium, while it is only the epithelium of the maternal vessel that has come in contact with them. And finally the third, that the foetal villi float in the blood of the lacuna, whereas they are always separated from it, by the wall of the maternal vessel and by the epithelium round it. Setting aside the deceptive appearances, it is evident that the relation between the absorbent foetal villi and the epithelium of the secretory villus is identical with that noticed in plans 9–10 and 11–12, the only difference consisting in the ectasia of the vessel of the maternal villus. Finally, at Nos. 15–16 is represented in diagram the human placenta. At first sight we perceive, on the right of the figure, that the phenomena apparent in the diagram designed upon the plan of the placenta of the dog are also met with in the plan

UNITY OF TYPE IN PLACENTAL DEVELOPMENT. 249

of the human placenta; but the truth cannot be held as demonstrated, even by an exact resemblance, which in reality rests upon an imaginary induction, and in treating of the human placenta this point deserves to be more minutely studied.

It is known that, in the earliest period of pregnancy, the placenta in the human species is represented by a layer formed of a rather compact cellular mass, called decidua serotina or placental decidua, in the midst of which runs a rich net-work of capillary vessels. From the surface of the chorion which is in contact with this, villi are formed, at first simple, which penetrate among the cells of the serotina. For a time the relation between these two parts of the placenta is not so close that they cannot be separated. It is known, moreover, that with the progress of the development, the vessels of the primary capillary net become ectasic and the primary simple villi proliferate and become arborescent.

If we could loosen and skillfully separate without tearing the compact mass formed by the cells of the decidua serotina (the placenta cells) before the vessels become ectasic, we should obtain exactly what comparative anatomy has very clearly shown us in the single placenta of certain animals, especially that of the dog, namely, a large-meshed netting of placental vessels, everywhere surrounded by a cellular layer, or the cells of the decidua serotina. The villi of the chorion which in the early period of development were, as we know, in contact with those cells, although they have penetrated into the serotina, would, if this were disentangled, be found actually in contact with the cellular envelope that clothes the vessels of the capillary net-work, and there would at once be repeated precisely what I have represented in diagram, at Nos. 9-10, which corresponds to the actual structure observed in the placenta of the cavia, as drawn from nature, in Plate IV., Fig. 2.

It was long ago known that, with the progress of development, the fœtal villi in the placenta of woman proliferate and become arborescent, and that, on the other hand, the net-work of the maternal vessels dilates and becomes ectasic. On this point the eminent Charles Robin taught[1] that, "in proportion as the villi of the chorion increase in volume, and are sub-divided, the superficial capillaries of the placenta are largely dilated and form delicate vascular folds interposed between the villi, which are still short, and in contact with the chorion as far as their peduncle, and that as the villi of the chorion multiply their divisions, so the interposed capillaries increase their dilatations." It results from this double series of facts, so well described by Robin, that in a compact mass, like that of the decidua serotina or placental, the new branches proliferating from the primitive villi are necessarily forced against the cells of the decidua and the walls of the maternal vessels. With the increase in number and volume, the cells and the walls of the vessels to which they adhere are introflected in the cavity which is formed in these vessels by their dilatation. This afterwards, in the complete development of the placenta, appears in lacunose form, but by the process now indicated the villus remains al-

[1] Mémoire sur les Modifications de la Muqueuse uterine pendant et après la Grossesse. Paris, 1861. Mémoire sur la Structure intime de la Vésicule ombilicale et de l'Allantoide chez l'Embrion humain. Idem.

ways covered and in contact with the layer of parietal cells that primarily clothed the maternal vessel. I have pointed out the succession of these facts at (o) No. 15 in the diagram.

In the human species when the placenta has been fully formed there is not the least trace of the primitive net-work of the maternal vessels, and the large arborescent masses of the fœtal villi may be said to swim in the maternal blood in so many compartments called cotyledonous cavities. For this reason Kölliker says [1] that if we imagine the chorion and all the villi taken away and only the uterine portion and the septa of the cotyledons left, the maternal placenta would have almost the appearance of a bee-hive.

Two of these cavities, or two actual cotyledons, are drawn in diagram at C (Nos. 15–16), indicating the formative beginning of a cotyledon, D a cotyledon in complete development. In this last, the envelopes of the parietal cells of the ectasic vessel have been forced against the chorion, to which they adhere, thus constituting the lamina of the sub-chorial decidua (i). The closeness and the union of the cells of the walls of two or more neighboring ectasic vessels form the septa of the cotyledons (l). The apertures or perforations in the septa which place in communication the blood of two so-called lacunæ, and the observations of Virchow, Kirvisch, and many others, who have seen the arteries communicate directly with the veins in the interior of the placenta by means of large openings, attest the permanence of that original vascular net-work, every trace of which

[1] Entwickelungsgeschichte des Menschen und der höheren Thiere, page 337. Leipzig, 1876.

has been supposed to be lost in the fully developed placenta, which does remain, however, deformed by the extraordinary ectasia it has undergone, and by the introflexion of its own walls determined by the proliferation of the villi. Hence it is an illusion that the fœtal villi float in the maternal blood, as it is an illusion that the epithelium that covers them appertains to the fœtus.

Recent observations do not materially change my former opinion, when, with Schroeder van der Kolk, I believed the cellular envelope of the fœtal villi to be furnished them by the maternal portion, and not to be their own epithelium, as is even now asserted by Kölliker.[1] At the beginning of the development of the placenta in woman, the villi have their own proper epithelium, which remains to them as long as they preserve a simple relation of contact with the cells of the decidua, but which they lose when the relation is intimately established in the manner before pointed out, and as we have seen it take place in cases of single placenta in all the mammalia. In short, in the epithelium of the villi of the human species we observe that to take place successively which is seen as separate and permanent in the different placentæ of the mammiferous animals. The epithelium is preserved so long as there exists a simple relation of contact between the maternal and the fœtal portion, and is lost when the relation becomes intimate; and it is singular that no anatomist had noticed the considerable differences which exist between the epithelium of the primitive villi and that supposed by them to clothe the villi when the placenta has completed its

[1] Op. cit., p. 333.

development, and when they describe the villi as swimming in the maternal blood.

The relation of the fœtal villus without epithelium to the secretory epithelium of the maternal villus is maintained in the human species also, thereby confirming the physiological idea elsewhere mentioned by me relative to the nutritive process of the human fœtus in the uterus of the mother. Late observations change nothing except the interpretation given as to the manner in which the cells of the decidua come in contact with the fœtal villi; and without the proofs now drawn from comparative anatomy regarding the origin of the cells of the decidua, the compact mass formed by them appeared to show that they took their origin from the connective elements of the uterus and were gradually transformed. With this conviction it was natural to believe that by means of the simple ectasia of the vessels the cells of the decidua might be carried over against the villi and enwrap them.

The crowded and compact mass of the cells of the placental decidua, traversed by a net-work of capillaries, completely obscured the perception of their origin from the walls of the placental vessels, a fact which comparative anatomy has to-day clearly established. Without knowing this, there was not, and there could not be, any knowledge of the exact structure of the human placenta, which, I think, may now be obtained with some simplicity and much clearness by adding what has been learned concerning this origin of the cells to the facts already possessed about the ectasic process that invades the net-work of the capillaries of the decidua serotina and placenta, and the proliferation at a determinate point and area of a single stem, or of a simple villus that becomes an arborescent trunk.

Without knowledge of these things, and before ascertaining the constant fact of the relation that is established between the fœtal villus and the cellular envelope that clothes the vessels of the maternal villus when the placenta is single, the genetic process of the human placenta could not be understood. It never was suspected, and the illusive appearances of the villi swimming in the maternal blood, and possessing an epithelium of their own, were held as demonstrated facts.

It should, however, be mentioned that Robin, as early as 1861,[1] had noticed the analogy existing in the relations between the maternal and the fœtal portions in the first periods of development of the placenta in the human species with those permanently established in the placenta of certain mammals. I hope it will meet the approval of that eminent scientist if, following the way luminously marked out by him, I have succeeded in proving that those first relations in the placenta of woman do not change even when the placenta has completed its internal development, but are identical with those met with in the single placentæ of all the mammalia.

This knowledge also explains certain conditions really existent and observable in the interior of the placenta, but which had no reasonable interpretation. I have shown that, reducing to its most elementary

---

[1] Mémoire sur la Structure intime de la Vésicule ombilicale et de l'Allantoide chez l'Embrion humain, dans le Journal de la Physiologie de l'Homme et des Animaux, page 334. Paris, 1861.

simplicity the intimate structure of a cotyledon of the human placenta, and to a single maternal vessel the vascular net that passes among the cells of the serotina, it must necessarily happen, by the very fact of the ectasia alone, that the decidual cells clothing the apex of the vessel should be carried against the chorion (Plate V., Nos. 15, 16, *i*); that these, placed along its longitudinal axis, must come in contact with corresponding ones in the neighboring vessels (*l*) forming the so-called septa, and that at length, the ectasia having become much more considerable, the cells placed at the base of the vessel must be carried over upon the uterine surface. Not aware of these very simple facts, Winkler imagined the existence of a species of decidual leaves or laminæ within the placenta. He gave the name of leaf or lamina of closure (*schlussplatte*) to the layer of cells found under the chorion, and that of fundamental lamina or leaf (*basalplatte*) to the layer of cells touching the uterus, and Kölliker[1] accepts these useless distinctions, but wishes to call the first one *decidua placentalis sub-chorialis*, and the other *decidua placentalis sensu strictiori*.

Nor are these imaginary leaves or laminæ enough. Some trunks of villi reach from the chorion to the uterine surface and there necessarily the wall of the maternal vessel is doubled and placed in contact with the cellular envelope of the vascular net, which is in direct relation with the uterine surface (*m*), and at this point there results a kind of collection of placental cells, which I described in my first work on the placenta in 1868, in order to show that the placental

[1] Op. cit., p. 337.

cells were continued over the villi. Two years later, this same observation was repeated by Langhaus[1] and lately Kölliker[2] attributes to him the merit of this discovery, which he judged interesting because he thinks it proves that the union of the two parts of the placenta is much more intimate than had hitherto been believed! and he proposes to call these extremities of the villi by the special name of roots of attachment (*haftwurzeln*).[3]

In the places where, through the ectasia of the maternal vessels, their walls come into a strict relation of contact, are formed by this alone, as I have stated, the so-called septa of the cotyledons, which Kölliker manages to divide into two layers,[4] and the apertures which are observed in these, and, as is believed, place in communication the blood of the alleged lacunæ, remain to represent the old capillary net-work of the primitive decidua.

It is only by insisting upon the mode of formation heretofore indicated that we can acquire exact knowledge of the intimate structure of the human placenta, and correct the fundamental errors hitherto prevailing that the villi possess a proper epithelium, and that they swim freely in the maternal blood by the formation of the lacunæ. From what has been demonstrated we must conclude that in the human species, as in all the mammals where the placenta is single, whether it be of zonarial or discoidal form, the fœtal villus without a proper epithelium, or having lost it, if there were such, at an early period of devel-

---

[1] Zur Kentniss der menschlichen Placenta, Centralblatt, No. 30, 1870.
[2] Op. cit., p. 336. [3] Op. cit., p. 333.
[4] Op. cit., p. 336.

opment, as is the case in woman, always comes in direct contact with the secretory epithelium which clothes the maternal villus. It is therefore only the introflexion of the walls of the maternal vessels upon the villi, and their enormous dilatation, that constitutes in certain quadrumana, and in the human species, a difference of form, though none of intimate structure, between their placentæ and those of all the mammalia having the placenta single.

It is well known that the ablest anatomists, among whom I may name Hunter, Bischoff, Weber, and Eschricht, not to mention others referred to in my first work on the placenta, were perplexed by the fact of the presence of the fœtal villi in the midst of the maternal blood contained in the lacunæ. Kölliker, not finding any of the opinions previously taught tenable, has recently made the assertion [1] " that the presence of the free villi within the maternal vessels can in no other way take place than by the increase of the tufts of the fœtal villi; the neighboring portions of the maternal tissue of the placenta and of the walls of the capillary vessels being thus compressed and destroyed, the lacunæ are therefore formed." Actual observations do not correspond to this supposed formation, but show in the placentæ of early abortions the gradual development of the lacunæ by ectasiæ from the beginning in the net-work of the capillaries amidst the mass of the placental cells, and while the villi of the chorion are still quite simple, as Robin has described with great accuracy. The formation of the lacunæ, therefore, precedes the formation of the tufts of the villi, and cannot be an effect,

[1] Op. cit., p. 341.

since observed before the existence of the cause assigned by Kölliker. But I have been able to make an observation much more conclusive on the formation of the lacunæ independent of the presence of the fœtal villi, having had occasion to study and describe the structure of the so-called uterine decidua in cases of extra-uterine pregnancy, and there found the exact and isolated anatomical structure of the maternal portion of the placenta in which existed lacunæ through ectasia of the vessels, without, of course, any trace of fœtal villi.

But it is not merely the inconsistency of the different opinions which have been broached for explaining the entrance of the fœtal villi into the lacunæ of the human placenta that helps us to the belief, based upon positive observations of comparative anatomy, that the relations between the two parts of the placenta are identical and constant in all the mammalia where the placenta is single, and, as I have said, of obscure glandular character. Other facts clearly observed in the human placenta, and demonstrated by men whose ideas are diametrically opposite to those which I have undertaken to prove, confirm the truth of my assertion, that the ectasia of the maternal vessels and the introflexion of their walls upon the villi constitute the only perceptible difference between the human placenta and that of the mammifera where it is single, and even in these the traces of ectasia are not, in some instances, wanting.

Among the facts indicated, I shall mention those observations having reference to the transformation of the uterine vessels when they become placental, and those directed to the so-called double epithelium of the villi. It has long been known that in the gravid uterus of woman are seen numerous small arteries with a spiral twist, of which arteries every trace is lost at the utero-placental limit and in the interior of the placenta.

Farre formerly imagined that these said arteries might open directly into the placental lacunæ. Kölliker has now taught that the short utero-placental arteries when they penetrate into the placenta lose their distinctive anatomical characteristics, that is, they no longer have muscular fibres or elastic elements, and their whole wall is formed by an endothelial layer, covered with a thin sheath of connective tissue, which disappears and blends with the decidua serotina. Likewise the veins are no longer to be distinguished from the arteries, and all traces of both are lost in the interior of the placenta, where alone the large lacunæ are found.[1]

Recently De Sinèty[2] has drawn attention to an important demonstration, which completes those made by Kölliker. He noticed that the cells of the decidua in woman form a circular sheath about the placental vessels, thus bringing a direct observation upon the human species into agreement with facts demonstrated with greater ease and certainty in some of the mammalia, and which I have described in the rabbit and the dog. Unhappily, De Sinèty shows that he had not at all understood the ideas and the facts relating to the placenta, set forth by me in my previous works, and very inappropriately quoted by him. However,

[1] Op. cit., p. 339.
[2] Archives de Physiologie normale et pathologique, VIII. An., p. 345. Paris, 1876.

if the facts observed by him in the completely developed human placenta harmonize with those already noticed early in its development, when there is seen a minute net-work of capillaries that become ectasic in the midst of the cells of the decidua or placental serotina, we shall be convinced that it is not that the walls of the utero-placental vessels are lost in these cells, as was indicated by Kölliker, but that so enormous a dilatation has taken place as to render it hard to perceive the endothelium, and to separate it from the placental cells that are elaborated by their external walls. Positive demonstration of this will be obtained by bringing together certain accurate observations which prove the endothelial wall of the vessels and their external cellular layer to be resting upon the fœtal villi.

Many anatomists and histologists used to affirm that in some portions of the villi they had observed two layers in the epithelium that covers the villi, and Kölliker has recently published[1] that, more especially when the epithelium is of suitable thickness, he has succeeded in distinguishing in it a deep layer with nucleated cells and a very thin surface, which, by means of acetic acid, may be detached under the form of a uniform membrane, in which the cellular limits are not very easily ascertained. The signification of these two layers, which, from the statements I have made, acquire no small importance, has not been brought out by any one, not even by Kölliker, who, with much ability, has described the change in the histological elements of the vessels, in their transition from uterine to placental. Even the observation repeated

[1] Op. cit., p. 334.

by him on the double epithelium led him only to remark that this second layer of epithelium of the villi had been frequently mentioned, and a little afterwards (holding only to the fact that in the extremity of the villi the epithelium appears single) he makes use of it to draw an argument in support of his opinion upon the mode of nutrition of the human fœtus during gestation. " In the extremities of the villi, at least," he writes,[1] " the fœtal vessels are, so to speak, immediately beneath the epithelium, and as the capillaries of the villi have in this place only the endothelial typical wall, and the epithelial layer which there covers them is itself very thin, and since they swim in the maternal blood, the passage of the maternal fluid into the interior of the fœtal capillaries must result without any difficulty." From this it would appear that within the human placenta the osmotic process for the nutrition of the fœtus would be easily carried on in the distal extremities of the trunks of the villi, but with difficulty in the villi, which are covered with the two layers of epithelium, and with still greater difficulty in the trunks, where abundant elements of the chorion surround the vascular loops and the minute net-work formed by them in the interior of the villi. Nor have the so-called epithelial buds of the villi, observed by so many anatomists in their different forms in the human placenta, hitherto had any signification whatever, but remained a simple recognition of fact.

I have elsewhere pointed out and brought together observations to prove that the so-called buds of the villi were only partial hyperplasiæ of the cells of the serotina which covered them, and within which pro-

[1] Op. cit., p. 335.

truded an internal vascular loop of the villus, and it was owing to this that the branches of the primitive trunk remained always clothed in a cellular envelope throughout their manifold divisions. Modern opinions do not change the ancient idea, but determine the fact with greater precision, especially since the recent investigations which I have instituted upon the intimate structure of the villi in the human placenta. In Plate IV., Fig. 4, are represented the results obtained by these investigations. The means employed by me consisted in the immersion for some time, in an extremely diluted solution of nitrate of silver, of portions of villi from a fresh placenta, then treating them with acetic acid before putting them in glycerine, or treating them with nitrate of silver and submitting them to the ordinary methods of coloring with carmine. With this and many other methods of treatment used by observers the double wall in the villi has been clearly demonstrated. In our illustration, and when the vessels of the villi are full of blood, they (a) are readily seen to be surrounded with the chorial tissue, in the midst of which appear large oval and granular nuclei (b). The chorial tissue and the fœtal vessels form the internal mass of the villus, which is everywhere surrounded with a cellular layer (c, c') as this is, in its turn, with an external membrane that appears like a homogeneous and transparent layer (d'). This is best seen at the terminal extremities of the branches of the villi (d), exactly where Kölliker asserts that the so-called external epithelial layer is not visible, and that the deep-seated layer is simple and very subtile. Only seldom and with some difficulty is any nucleus seen in this **enveloping membrane**.

These observations of mine agree, then, at least in the fundamental part, with those made by Kölliker, and by many who have stated that the foetal villi of the human placenta are covered with two layers of epithelium. But how can that be called epithelium which is a uniform and compact membranous layer, in which, as Kölliker himself taught, the cellular elements are only with great difficulty discovered?

With regard to the nucleated cells of the deep-seated layer, as it is called, all are agreed. Now, if we mentally picture the removal, from the tissue of the chorion and from the foetal vessels of a villus, of this double envelope by turning it back, do we not have a maternal placental vessel formed of a single wall everywhere surrounded by decidual cells, as we have observed in the placenta of many mammals having the placenta single? Have we not, in short, also in woman the elementary typical form of the maternal villus?

But, confining ourselves to the scrupulous examination of fact, in the figure in question, at g is indicated the process by means of which occurs the ramification of a trunk of a villus. The protrusion of a vascular loop, which shall form a new branch of a villus, against the inner wall of the ectasic vessel occasions a new introflexion within its cavity (a supposed lacuna), not only of the internal cellular wall, but also of its external wall, in which the structure of the endothelial cells is either much modified, or we have not been able clearly to demonstrate it.

The mammillary shape of the epithelial buds of the villi, as is perceived at the apex of the branch of the villus in development (g), and the pedunculate form of the buds indicated at *h*, which are so frequently met with in the villi of the human placenta, very probably show, the first, an initial and progressive process in the formation of a branch of a villus, and the second, an arrest of development in that process.

The accurate knowledge now presented upon the formative process of the placenta, upon the intimate structure of the new-formed placental vessels, upon the origin of the decidual and placental cells, as well as upon the ectasia of the maternal vessels in the placenta of the quadrumana and of woman, appears to me to be all in harmony, one part with another. The different observations that have been gradually accumulating, upon which a final decisive opinion was still wanting, receive a clear and simple interpretation, and I do not believe myself mistaken in affirming that the ideas acquired substitute simplicity and clearness for the many uncertainties and important errors which have hitherto been held as to the anatomical structure of the human placenta. By these researches we have moreover been able to demonstrate that the relations between the fœtal and the maternal portion of the placenta in woman are perfectly identical with those observed in single placentæ in the other mammifera. We have traced amid the manifold and remarkable varieties of form the unity of the anatomical type of the placenta in the different classes of mammiferous animals, and discovered, as resulting from it, the physiological unity of the law that governs the nutrition of the fœtus in all vertebrates, whether its life is completed within the body of the mother or in an egg outside of it.

In summing up, then, the subjects discussed, the

principal facts may, I think, be comprehended in the following conclusions : —

(1.) Immediately after conception, the first fact which is established in the uterus is a destructive process that affects its inner surface. In some animals and in woman this process is limited to the mere epithelium, whilst in other animals, as in the rodents, the destructive process attacks the whole sub-mucous connective layer, including the vessels, nerves, and utricular glands found therein, as far as the inner surface of the uterine muscular tissue.

These two facts, which seem in regard to their importance so widely divergent, have nevertheless, in the rodents as well as in woman, one common, ultimate, and identical result, namely: that of the denudation of the internal surface of the uterus.

(2.) The extent and depth reached by the destructive process in the full segments of the gravid uterus in rodents serve to show, very plainly and positively, that the formation of the decidua and placenta is due, neither to a tumefaction nor to a transformation of the anatomical elements preëxistent at the time of conception in the uterine mucous membrane; and that, if the utricular glands also are affected and destroyed before the placenta is developed, they are necessarily to be regarded as completely extraneous to the formation of the placenta. Though in other animals, when these remain increased in volume through the whole time of gestation, I cannot deny to them some share in the nutrition of the fœtus, still the facts observed in the rodents prove that they are not essential to the nutrition of the fœtus in all the mammalia.

(3.) The destructive process more or less deep on the internal surface of the uterus is in all cases indispensable, because this is what facilitates the setting up of the neo-formative changes from which will result the maternal portion of the placenta. The deep-seated destructive process which is observed in the rodents serves admirably to demonstrate the neo-formation of the decidua and placenta, which in the rat has been followed in its minutest particulars from the moment of the arrest of the ovum in the uterus.

(4.) The neo-formative process of the uterine or maternal portion of the placenta consists in the production of new vessels, which are distinguished from the ordinary uterine vessels by two special characteristics: first, the arterial as well as the venous vessels have only a simple endothelial wall; second, from the external surface of this wall is elaborated a layer, more or less thick, of special cells not separable from the wall of the vessel. These are the so-called decidual and placental cells. That the fœtal portion of the placenta is itself due to a neo-formative process has not been, and cannot be questioned.

(5.) It is from the constant relation established between these two parts of new formation that the placenta is developed. The manner in which this relation is established gives rise to the different forms of placenta known in the mammalia.

(6.) The elementary or primordial form of the two fundamental parts, from which results the placental formation, may be anatomically determined by the form and structure of a simple villus, by the fœtal as well as by the maternal portion; the function alone differs in the two villi, being absorbent in the fœtal, and secretory in the maternal villus.

This elementary and typical form of the two parts constituting the placenta in the mammifera is not imaginary, but is demonstrated in its simplicity by observation. It is easily enough recognized in the fœtal portion, in the villi of the chorion, especially in the simpler forms of diffused placenta, as, for instance, in the sow and in the cetacea, and at the beginning of development in the human species. In the maternal portion the simple, elementary form is found to be developed and maintained through the whole period of gestation in the uterus of certain viviparous fishes.

(7.) The manner in which the relation between the two parts is established may be by simple proximity, contact, or by intimate cohesion. When the relation is that of simple nearness, the maternal portion of the placenta manifestly presents the form of a glandular organ and has its limitation by the repetition of secretory villi upon the inner surface of the uterus, which, uniting with each other in various ways, give rise to the formation of crypts or glandular follicles, simple or compound, into which enter the absorbent villi of the chorion. When the relation is more intimate and an adherence takes place between the two parts before mentioned, as in cases of single placenta, whether of zonarial or discoidal form, the glandular character is concealed by the very fact of the adhesion, but the fundamental condition remains constant, the contact, in this case direct, between the vessel of the absorbent villus and the epithelium of the secretory villus, which is never lost in any instance.

(8.) Only two very simple changes occur in the

fundamental parts of the placenta when single, and they are the factors of the manifold differences which are met with : first, the loss of the epithelium of the absorbent villus, which is not important, since there is established direct contact of the vessel of the villus with the secretory epithelium of the maternal villus, and this fact is constant; second, the dilatation or ectasia of the vessel in the maternal villus, and this fact is remarkable only in the placentæ of the quadrumana and of woman. The ectasia in the maternal vessels, already shown under a rudimental form in the placentæ of certain mammals, has been indicated by Eschricht and by Turner as representing the large lacunæ which are observed in the placenta of some of the quadrumana and of the human species. But these supposed lacunæ have been the chief, if not the only stumbling-block to the exact knowledge of the structure of the human placenta, although anatomists were aware that ectasia took place in the placental vessels in their first phases of development. The belief that the lacunæ were really large cavities, as they had the microscopic appearance of being, and not the maternal vessels greatly dilated, was held by all as a truth proved and indisputable, and it was through this belief that two other deceptive appearances were received as actual truths, namely: that the villi floated in the maternal blood, and that the epithelium covering them appertained to the fœtus instead of the mother.

I have demonstrated as facts certain things ascertained with regard to the earliest period in the development of the human placenta, namely, first, that the maternal vessels which course in the compact

UNITY OF TYPE IN PLACENTAL DEVELOPMENT. 269

mass of the decidual cells and the placenta are affected by an ectasic process; second, that the foetal villi, when simple, are in contact with the aforesaid cells; third, that the simple foetal villi at first proliferate and then become arborescent. I have shown that, harmonizing these facts in their progressive development, there must of necessity occur a protrusion of the villi within the cavity of the dilating vessels, and that this protrusion cannot take place without an introflexion of the walls of the vessels that are pressed upon by the proliferating chorial villi; therefore the first relation that is established in the human placenta between the chorial villi and the cells of the decidua is still maintained unchanged, even when the placenta has completed its development. As a consequence of the intimate relation existing between the foetal and maternal portions, the villi lose only their primitive epithelium, and finally there is also established in the human species, between the two parts of the placenta, a relation identical with that which we have observed in single placentæ in other mammals in which is wanting only the ectasia of the maternal vessels.

Setting aside the fundamental errors, many facts already observed by able anatomists in the human placenta, which have remained doubtful or were wrongly interpreted, now receive a clear and precise explanation.

(9.) The belief that the villi in the placenta of woman were floating in the blood of the lacunæ generated the physiological error that the nutrition of the foetus, not only in the human species but in all mammals, took place through an osmotic exchange of the two bloods, although in the case of diffused or multi-cotyledonal placentæ, the facts openly contradicted such an assertion. Besides, in all cases, where the placenta is single, the vessel of the absorbent villus (of the fœtal portion) never comes in contact either with the blood or with the wall of the maternal vessel, but there is always interposed between the walls of the two vessels, and consequently of the two bloods, a cellular layer which is the epithelium of the maternal villus; and that this is secretory is confirmed by the obvious glandular appearance which is observed in many animals in the maternal portion of the placenta when it has the diffused or multi-cotyledonal form.

(10.) The nutritive material which is to serve for the growth of the fœtus in all the vertebrates is furnished by the mother. In mammals it is supplied by the maternal portion of the placenta, gradually, as the fœtuses are developed. In the oviparous vertebrates the material, in the quantity necessary to the development, is emitted in a mass from the mother, in the form of yolk with the egg. In the mammiferous, as in the oviparous, animals the absorbent or fœtal part of the placenta does not change, and it is by means of an absorbent villus, more or less complicated, that the material elaborated by the mother is conveyed to the fœtus. It is, therefore, but one law, a physiological modality, that governs the nutrition of the fœtus in all the vertebrates.

(11.) The observations made upon the primordial or rudimentary forms of the placenta in certain viviparous fishes leave it doubtful if the marsupialia should be considered as mammals without placenta. From the very little known about it, even in these animals, there exists, for the nutrition of the fœtus in the body of the mother, the relation of contact of a secretory surface, the uterine, with an absorbent surface, fœtal chorion without villi. This relation is observed, in *Mustelus lævis*, between the large umbilical vesicle without villi and the large folds of the uterine mucous membrane, which embrace and surround the fœtuses and their envelopes. The exact actual observations which have been also lately collected by Turner in no wise, then, allow us to accept the recent opinions of Kölliker, who considers all the mammals of diffused placenta as implacentate.

SUMMARY AND CLASSIFICATION.[1]

IN the ten years that have elapsed since the publication of my first inquiries into the intimate structure of the placenta in the mammals and the human species, I have not let pass a single favorable opportunity of procuring new material for study, either to confirm the observations already made, or to extend and enlarge as much as possible the opinions I had published, and to correct them if erroneous.

But the great difficulty met with in obtaining gravid females of animals at different periods of gestation renders the observations less complete than I could wish, and the poverty of those who in Italy cultivate the natural sciences prevented me from making any investigation or acquisition abroad, as I very much desired, and my work would have remained isolated, certainly it would have borne little fruit, but fortunately during this period the distinguished anatomist,

---

[1] Professor Ercolani writes me, in reply to inquiries as to the conclusions at which he has arrived, since the publication of his memoir, that while gratified that the "results of his long-continued studies are about to be presented to the knowledge of his colleagues in the great American republic," he confesses a hesitancy in offering further deductions because of the fear that his "long-continued labors might have led to conclusions so important, that in announcing them it would be a duty to furnish a long series of actual observations, with the inferences and deductions drawn from them." Since opportunity is not afforded for doing this, he has kindly sent me what he considers "a brief and imperfect summary of the labor thus far accomplished." I take great pleasure in adding this to the present work, and believe it will be accepted by scientists as of great value. — H. O. M.

Professor William Turner, of Edinburgh, devoted himself, with a rare steadiness of purpose, to this kind of investigation, and had the happy opportunity of bestowing his learned and conscientious researches upon no small number of animals either rare in themselves or difficult to obtain in a state of pregnancy. He availed himself of the rich material for study which cultivated England can afford to her sons who honorably pursue the different branches of human knowledge. The studies and the numerous observations of Turner, and those more limited and fewer of Harting and Creighton, gave me the greatest satisfaction, because they either confirmed those I had made, or, by extending them to other animals which I had not examined, largely established my fundamental demonstrations.

At first there appeared a slight disagreement between the observations of Professor Turner and my own. I had very concisely asserted, contrary to the doctrine taught by Sharpey and received by other able anatomists, that the utricular glands of the uterus not only did not receive in their interior the villi of the chorion, to form with them the placenta, but that they did not at all enter in any case into the formation of that organ, and had no important office in the nutrition of the foetus, since it, by means of its envelope, was placed in direct relation with the uterus of the mother. Turner, beginning his researches on the placental formation of the *Orca gladiator*, showed himself doubtful about accepting this general conclusion, so remote from the ideas then held by the most famous anatomists, but afterward extending his investigations to other animals having a diffused placenta, as has the orca, and to those where it is pluri-cotyledonal in form, as in the domestic herbivora, with noble sincerity he repeatedly declared that the utricular glands in no case took part in the formation of the placenta, and that the uterine crypts which receive the villi, even in the *Orca gladiator*, are to be regarded as altogether of new formation.

Apart from this slight disagreement, which, as regards the manner of nutrition of the fœtus in the uterus, was of secondary importance, the fact still remained, which had seemed to me of greater weight, and which I had sought to bring out clearly, namely, that in animals as well as in woman the nutrition of the fœtus is in no case effected by means of an osmotic exchange between the two bloods, but always takes place by the neo-formation of a special secreting organ, furnished by the uterus of the mother. Turner, for the designation of glandular organ given by me to the secretory organ, substituted the more general and comprehensive name of secretory surface, which is either placed in simple contact, or else intimately united, with an absorbent surface appertaining to the fœtus.

The exactness and precision of the observations of Turner and others thus strengthened my own conclusions, and with a generosity rather unique than rare, valuable material for study was placed at my disposal by the distinguished professors Theodore Bischoff and Milne Edwards, to whom I am happy to express the sincerest gratitude. By means of the careful investigations I was thus enabled to institute, there was afforded me a good opportunity to enlarge the sphere of my views, and to extend the narrow circle of the analysis and comparison of separate facts, to which for a long time I had been obliged to confine myself.

The new facts observed, though adding some beautiful and interesting characteristics useful to be known, all relate especially to the manner of the structure of the placenta in different animals, and, so far from changing, they rather confirm all the deductions and the fundamental opinions which I had stated in my earlier works.

To describe the anatomical peculiarities seen in the intimate structure of the placenta in those animals recently examined is not, I think, what you require of me, nor would this be the fit place for making a minute and necessarily long exposition of them. I flatter myself, therefore, that it will be more satisfactory to sum up briefly all the anatomical facts relative to the most remarkable differences observed in the placentæ of the different mammifera, and then to touch on the general conclusions which, in my opinion, are to be drawn, applying what has been learned to the taxonomy and the phylogeny of the mammals, subjects on which learned scientists spend, and have spent, many years of labor.

While carrying on investigations upon the formative process of the placenta in woman and in the females of the mammiferous animals, I was led to trace out the unity of the anatomical type from the histological point of view in all the different forms of placenta which up to that time had been observed. I endeavored to show that this unity consisted in the neo-formation of a simple villus, in the strict anatomical sense of the word, on the part of the mother as well as on the part of the fœtus (villus secretory or maternal; villus absorbent or fœtal). I now add that to this single type are subordinated also the new forms of placenta which have been seen and described by others and by myself.

I was at that time content with having shown that to the unity of the physiological process or office of the placental organ corresponds the histological unity of the anatomical type governing all the various forms under which the placenta is observed; that they owe their origin solely to a different relation which is established between the maternal and the fœtal villi; a simple relation of proximity and contact in the diffused and many-cotyledoned. in which, whatever may be its form, the epithelium of the absorbent or fœtal villus comes in contact with the epithelium of the secretory or maternal villus; a relation of intimate union, on the other hand, between these two species of villi in the zonarial or single discoidal placenta. I also stated that there were but two very simple factors in these apparently very important anatomical differences; that they consisted in the loss of the epithelium in the absorbent villus when the simple relation of contact is changed for that of intimate union, and in the ectasia or non-ectasia of the vessels of the maternal villi when we inquire into the differences between the single discoidal placenta of animals and that of the human species.

Later observations have shown me that not all the zonarial placentæ have an identical intimate structure, as is generally believed, but that even in animals where it is single and of discoidal form there are examples of the duplex and differing relation between the fœtal and the maternal villi, or between the maternal and fœtal portions of the placenta. This diversity has served zoölogists as a basis for the division of mammals into deciduate and non-deciduate.

I think that these recent investigations may have some importance in disturbing the fundamental basis accepted by zoölogists for distinguishing the mammalia, and that they may also, with some simplicity and clearness, lead to the recognition of the fact that not only the histological anatomical type, but also the microscopic anatomical type, preserves a true unity of form. Proof of this is obtained by examination of the initial or rudimentary forms in which the placental organ manifests itself in certain cartilaginous fishes.

Müller was the first to distinguish, among cartilaginous fishes, the Plagiostomi into acotyledonous and cotyledonous, placing with the former those where the ova, after having been fecundated, remain up to their complete development in simple contact with the internal wall of the incubating chamber; and with the latter, those in which the ova, with a part of their large umbilical vesicle, become intimately united with the inner wall of the chamber or uterus, by means of mutual introflexion between the folds of the two surfaces. Later, Bruche described a villous neo-formation on the mucous membrane of the gravid uterus of certain species of fishes, the Selachii, which forms a sort of nest from which the ovum of those fishes derives the material necessary to the complete development of the fœtus.

All the principles that underlie exact knowledge of the intimate structure and real value of the manifold microscopic appearances which the placenta assumes in all mammals are included, as I believe, in these few and simple facts observed in certain fishes. For if in the past anatomists and zoölogists were confused and embarrassed by the varied and manifold forms of the placenta in the different mammiferous animals, to-day we are, I think, on the other hand, forced to admire the simplicity with which nature maintains and exhibits unity exactly where there appeared to exist the greatest and most remarkable diversity.

Uniting all the observations made by others or myself upon the exterior forms and intimate structure of the placenta, which permit being judged of with certainty (many older observations not admitting of such accuracy), I shall now attempt to show that in a comparison of form and structure in the placental organ in the mammalia, the same very simple facts are repeated, though more fully and perfectly, which I have already pointed out as seen in certain cartilaginous fishes. It will be noticed that I have indicated rather more forms of placenta than are generally recognized by anatomists, and, unfortunately, I am obliged in this communication to present as mere assertion that which needs to be demonstrated with a minute anatomical description of facts, to which I can here only allude.

I trust, however, that for those who have pursued carefully scientific investigations of this character even these hasty suggestions on the different placental forms and structure may suffice to convey my idea in explanation and completion of the teachings left us by the illustrious von Baer. The facts adduced by this learned embryologist to distinguish two fundamental forms of placental development in mammals, from conception to the act of delivery, with or without traumatic lesion of the internal surface of the uterus, remain unchanged, because they are indisputably true, and correspond exactly to those observed by Müller and others in the acotyledonous Plagiostomi without lesion of the chamber of incubation, and in the cotyledonous with lesion of the chamber by separation of the folds which were closely united during gestation. The distinction taught by von Baer occurs in certain cartilaginous fishes, and has its counterpart in that advanced by Huxley for the mammalia, the acotyledonous Plagiostomi corresponding to the non-deciduate, and the cotyledonous to the deciduate mammals. Harmonizing these facts with what is known of the placental organ, the doctrine of von Baer may be thus expressed : —

(1.) Mammalia in which the nutrition of the ovum takes place through simple contact of its absorbent surface with the internal or secretory surface of the uterus. (Acotyledonous Plagiostomi.)

(2.) Mammalia in which the nutrition of the ovum takes place through the intimate contact and union of a part of the absorbent surface of the ovum with a part of the internal or secretory surface of the uterus. (Cotyledonous Plagiostomi.)

Let us now see what are the successive changes, or modifications, which, still maintaining the fundamental typical form seen in the fishes, are met with in the placentæ of those of the mammalia that have hitherto been observed and studied.

ACOTYLEDONOUS PLACENTÆ.

Under this denomination I include all those of placental formation observed in mammals in which, notwithstanding the numerous and varied microscopic forms they present, there is maintained in all cases propinquity or simple contact between the absorbent or fœtal and the secretory or maternal surfaces. As to the exterior forms the following species may be distinguished.

ACOTYLEDONOUS PLACENTÆ, SIMPLE.

This form, which, so far as is known, is that most nearly resembling that of the acotyledonous Plagiostomi, may have been observed in the Ornithodelphia in *Ornithorhyncus paradoxus*, and among the Didelphia in *Macropus major*, but it must not be concealed that positive observations are still wanting to prove that in this last, at least, there may not take place on the internal surface of the uterus a villous neo-formation analogous to, or representing, that found in some fishes among the Selachii.

ACOTYLEDONOUS PLACENTÆ, VILLOUS AND DIFFUSED.

In this form of placenta the relation of simple contact between the two surfaces of absorption and secretion is greatly increased by numerous foldings of both surfaces, and by the neo-formation of many villi upon them. In this group are included the following:

The simple diffused villous form in which the facts already indicated are confirmed; the very numerous villi of the two surfaces come in contact and are disposed in linear series or in groups on the folds of the chorion and the uterine mucous membrane. This form of placenta has been observed in *Sus scrofa*, *Propithecus Verrauxii*, *Lemur rufipes*, and in some species of the genera *Lepilemur*, *Hapalemur*, and *Cheirogaleus*.

The compound diffused villous form differs from the preceding only in the fact that the secretory or maternal villi, over the whole internal surface of the uterus, are united to each other in such a way as to constitute so many small depressions, having the shape and the office of glandular organs, which receive in their internal cavities the fœtal absorbent villi. According to the greater or less degree of perfection of the neo-formed glandular organ in this general type of placenta, we may make a distinction between those in which the glandular organ is of cryptal form and those where it is follicular. The cryptal form has been described in *Pangolinus*, *Delphinus phocœna*, *Balœnoptera Sibbaldii*, *Orca gladiator*, *Halicore dugong*, *Tragulus Stanleyanus* and *meminna*, and in *Hyomoschus aquaticus*. The rhinoceros and tapir families present this form. The follicular has been described in *Equus caballus* and *asinus*, in *Camelus dromedarius*, and in *Monodon monoceros*.

ACOTYLEDONOUS PLACENTÆ, COMPLICATED VILLOUS.

There is but a single example as yet known of this form of placenta, which has been observed only in *Elephas indicus*. It differs from those already mentioned in this: at the poles of the ovum it presents a simple and localized diffused villous form instead of at its middle part, where is developed a species of distinct zone, composed of villi closely crowded together and received into numerous new-formed glandular crypts.

ACOTYLEDONOUS PLACENTÆ, LOCALIZED VILLOUS.

The simple contact between the absorbent and the secretory surfaces is maintained, as in the foregoing forms of placenta, save that the relation of contact is more or less complicated and is established exclusively at certain points, which may be very numerous, few in number, or even at one place only, more or less extended or limited. The glandular neo-formation thus localized upon the maternal surface is perfected and elevated in its structure, acquiring the forms of a compound follicular gland. By reason of this complication the chorial villi are given off from the foetal surface in tufts, or groups of tufts, of villi more or less branching or arborescent.

I do not accept the name of pluri-cotyledonous, employed for these forms of placenta, because, as I shall point out, I have seen this structural type in placentæ of zonarial and even of single and discoidal forms. In this group, then, we may distinguish the following : —

With diffused and extended localization over many points of the two surfaces.

Examples of this are afforded by the well-known placental formation in *Bos taurus*, *Ovis aries*, and *Capra hircus*. Here, also, is to be referred the manner of forming the placenta in *Camelopardalis giraffa* and probably also that of *Hippopotamus amphibius*, in which, however, the centres of localization would be smaller and more numerous.

With localization limited to a few points only of the two surfaces.

In this form are classified the placentæ observed in *Rangifer tarandus* and in *Cervus dama, capreolus, elaphus, axis,* and *porcinus.* A remarkable peculiarity or exception relative to placental formation in the deer has lately been described in *Cervus mexicanus,* in the gravid uterus of which were observed three large cotyledons, several dozen smaller ones, and also numerous small, simply villous spots. In this is thus afforded a fine example of passage from the forms of simple villous placenta to the far higher one now under consideration.

With localization limited solely to the middle zone of the ovum.

To this form are to be referred the zonarial placentæ described in *Phoca vitulina, Phoca bicolor (Monacus albiventer), Halichœrus grypus, Hyrax capensis,* and among the Edentata in the genus *Orycteropus.* The same is met with also in *Lutra vulgaris* and in *Mustela foina, martes,* and *vulgaris.* I ought here to remark that the zonarial form which the placenta has in these animals has led anatomists to believe that therefore the intimate structure must also be identical with that of the zonarial placenta of the carnivora, which, as I am about to show, is entirely different.

With localization clearly distinct at one point of the surface of the ovum.

The placenta is single and of discoidal form. The intimate structure is that common to all acotyledonous placentæ. It has not hitherto been observed except in *Talpa europœa*.

COTYLEDONOUS PLACENTÆ.

I include under this generic name all those forms of placental structure which have been found in mammals where the fundamental character that distinguishes them from those previously indicated is never wanting, namely, the intimate union, at a space of greater or less extent, between the secretory and absorbent surfaces, a union to which I have already called attention as taking place by means of certain folds in the incubating chamber in the Plagiostomi, for this reason called by Müller cotyledonous.

The secretory surface in these placentæ is formed by the maternal vessels which convey in the maternal blood the materials necessary to the secretion, and by the organic secretory elements, which consist of the peri-vascular or deciduate cells that, in all cases, surround the maternal placental vessels. The fœtal or absorbent surface is also formed by the vessels which convey to the embryo the separate elements capable of assimilation. The chorial villi, constituting the absorbent surface, coming in contact and close union with the deciduate cells of the secreting surface, lose very properly their external epithelium, rendered by this union useless and troublesome to the villi for the discharge of the duty they are to perform; therefore these forms of placenta may be styled vascular, since the chief differences noticed among them depend upon the manner in which the union is effected between the two vascular surfaces.

The species of these placentæ are:—

Cotyledonous Placentæ of incomplete vascularization.

Even the older anatomists distinguished two parts in these forms of placenta, one which belongs to the uterine surface, and which the vessels of the foetal villi do not reach (the maternal portion of the placenta); the other which is carried over it and towards the uterine cavity and the foetus. With this part alone come in contact the vessels of the villi of the absorbent surface, which, deprived of external epithelium, unite with the secretory cellular elements that surround the vessels of the maternal villi (the foetal portion of the placenta). This form of placenta has been observed in *Cavia cobaya, Dasyprocta aguti, Mus musculus* and *decumanus*, and in *Lepus cuniculus* and *timidus*.

Cotyledonous Placentæ with complete vascularization.

The foetal vessels of the absorbent surface divested of epithelium come in direct contact with the secretory cellular elements which surround the vessels of the whole maternal portion. Ordinarily in this kind of placental formation the diameter of the maternal vessels does not show partial ectasic dilatations, and only in a few cases are there rudimentary indications of them anywhere. But dilatations in the form of lacunæ are always wanting, and there is no distinction of the placenta into two parts as in the preceding species. The shape of these placentæ is zonarial in some animals, in others discoidal. It is zonarial in *Canis vulpes* and *domesticus*, and in *Felis catus domesticus*. It is discoidal, and has been studied, in *Erinaceus europæus*, in *Vespertilio murinus* and *noc-*

tula, in *Pteropus medius, Noctilio leporinus,* and *Phyllostoma hastatum.* Very probably this same form has been observed in *Centetes ecaudatus.*

Cotyledonous Placentæ with complete vascularization and with ectasia in the vessels of the secretory surface, or maternal portion of the placenta.

In this form of placenta the external surface of the dilated vessels of the maternal part conveys the perivascular and secretory cells against the walls of the vessels of the absorbent fœtal surface, which, according to the degree of dilatation that takes place in the vessels, remains only in simple contact with these, or is co-involved with the secretory cells and with the walls of the maternal vessels. The external form of the placenta and the degree of dilatation of the uteroplacental vessels may serve to determine some species of placentæ in this group, also, *e. g.* : —

External form bell-shaped and composed of many lobes united together, with regular and uniform dilatation in the vessels of the secretory surface.

This variety of placenta is found in *Bradypus tridactylus* and *didactylus,* or *Cholœpus Hoffmanni;* in this appear, also, certain larger dilatations at some points in the said vessels, representing the beginning of the formation of lacunæ.

External form unilobed, discoidal, or circular, with irregular ectasic dilatation of the vessels of the secretory surface, so as to produce lacunœ.

Observed in *Tamandua tetradactyla,* and *Dasypus gymnurus, novemcinctus,* and *sexcinctus.* To this form is probably to be referred that seen in *Cyclothurus didactyla*.

External form bilobed, that is, composed of two masses of discoidal shape, more or less separate from each other. Lacunose ectasia in the maternal vessels.

Observed in different species of *Simiæ*, as in *Hapale jacchus, Mycetes ursinus* and *seniculus*, in *Cercopithecus sabeus, Cynocephalus sphinx, Semnopithecus nasica* and *mitratus*, and in *Macacus nemestrinus* and *cynomolgus*, etc.

External form unilobed discoidal. Lacunose ectasia as in the foregoing.

Observed in the Anthropoid apes, in *Troglodytes niger*, and in the Human species.

I am aware that some at least of the anatomical facts which I have simply indicated need to be minutely described in order to acquire such value as they may have, but it is impossible to do this in so brief a summary as is here given, and I therefore confine myself to the outline as furnished. I intend to make a full demonstration, with anatomical analysis of the facts, in order to show that there are animals having the placenta of zonarial and discoidal form which are not deciduate in the anatomo-zoölogical sense of the word as now employed by modern writers; and that the ectasia of the utero-placental vessels, even when presenting the appearance of large lacunæ in the interior of the placenta, not only is not an anthropological characteristic, as was once thought, of the human species, but besides being common to the *Simiæ* has been also observed in the placenta of some among the *Edentata*.

In considering the various forms noticed in the placenta of the different mammals, others had been successful in detecting a gradual and progressive development from the simplest up to the highest, those of the monkeys and the human species; and that this regular progress in the development of the placental organ bore no direct relation to the rank held by the different species of mammals in the scale of being. The careful researches and minute distinctions which I have above recorded help to confirm these very important conclusions. I call them very important, because they are to-day brought forward to show that inquiries concerning the form and structure of the placenta have not, and cannot have, any useful application to zoölogical taxonomy and the phylogeny of the mammalia.

I have already pointed out in preceding works, that I thought the basis accepted by zoölogists for distinguishing the mammalia into deciduate and non-deciduate to be erroneous, the uterine decidua representing in all these animals the initial neo-formative process which gives rise to transformations, more or less extensive, of the secretory surface, or, in other words, to modifications which the new-formed villi undergo upon that surface. In mammals styled non-deciduate, having diffused placentæ, whatever may be its species, the whole deciduæ or primitive villi are changed into secretory organs of more or less simple glandular forms, and the same fact is repeated, but limited to certain points only of the secretory surface, in non-deciduates having pluri-cotyledonous placentæ. In woman, the type of the deciduates, the decidua, indeed, is formed upon the whole internal surface of the uterus, but there is only one part of the new-formed decidua, that, namely, against which the ovum is fixed, that becomes placental; all the remaining portion is arrested in its development and becomes that external part of the fœtal envelope which is known as decidua vera, concerning the origin and structure of which there is so much discussion by anatomists and obstetricians.

According to these views the true deciduates *par excellence*, those, namely, in which the whole of the primitive decidua is transformed into maternal placenta, would be those mammals which are thought non-deciduate. If we restrict the name of deciduates to those in which, as was pointed out by Weber and von Baer, there takes place in the act of parturition a traumatic lesion of the uterus, through the caducity of the maternal portion of the placenta, even then the received basis is fallacious, since, as I have indicated, not only among animals having zonarial placentæ would some be deciduate and others non-deciduate, but there might be an example of one considered deciduate, with the placenta single and discoidal, in which in the act of delivery the two parts constituting the placenta become enucleated as happens in the cotyledons of some Ruminants.

That among the *Edentata* there were some that should be classed with the deciduates, and others with the non-deciduates, was already known to zoölogists, who considered these facts as exceptional and confined to that order of mammals. It appears to me, however, that the exception goes too far, and as such can no longer have any actual value.

Anatomists and zoölogists have always foreseen the

importance that studies on the placenta must have in a natural classification of the mammalia. The numerous attempts made from Everard Home down to Huxley show this unquestionably. But after so many toilsome and minute researches, anatomists and zoölogists have been led to conclusions entirely opposed to the end for which they had expended so much time and labor. As step by step they acquired more extended and complete knowledge of the form and intimate structure of the placenta, the conviction deepened that the doubts already raised by Owen as to the applicability of these studies to zoölogical taxonomy would not lead to useful results, and to-day, through the labors especially of Rolleston and Turner, it is plainly asserted that researches on the placenta neither have nor can have any useful application to zoölogy.

Is this assumed discrepancy between embryological and zoölogical studies to be held as sufficiently proved, and henceforth to be considered indisputable?

I will also, on this most important question, briefly state the conclusions at which I have arrived. Notwithstanding the serious disappointments sustained by all those who had hoped for a blaze of light from the application of studies upon the placenta to zoölogical taxonomy and phylogeny, I do not think that there were many who were willing to abandon a conviction which, though reached *a priori*, was yet deeply and generally felt by learned men, as the history of science shows. The idea appears too simple and too logical to be relinquished, that the form of the foetal envelopes and the placenta, which have so great a share in the nutrition of the fœtus, must have, in the embryological point of view, a nexus or closely connecting bond with the forms which the animals will exhibit when arrived at their complete development. Though we may be compelled to abandon even the hope that the paths hitherto followed for applying the knowledge obtained about the form and structure of the placenta to zoölogy and phylogeny may lead to useful results, yet it has seemed to me that recalling to notice and expanding, by the knowledge now possessed, an idea which was expressed by von Baer, but neither by him nor by others sufficiently considered, we may arrive at some important conclusions.

It was known to von Baer that in the fœtal envelopes of certain mammals, the umbilical vesicle increased in volume during the period of gestation, and remained voluminous even in the act of parturition; and upon this fact, with that of the rapid disappearance of the aforesaid vesicle in other animals, von Baer proposed to classify the mammals as follows: —

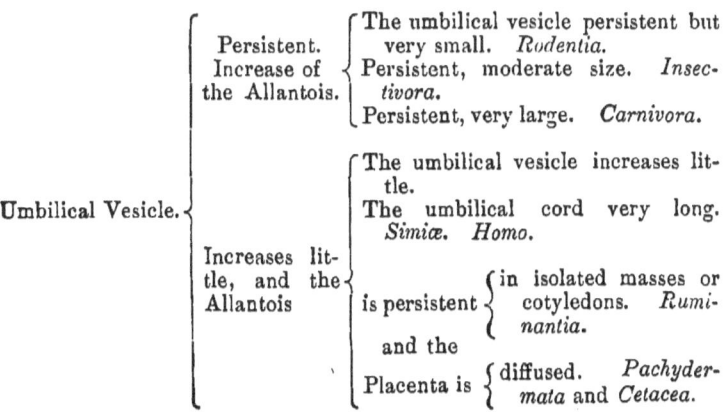

I shall not dwell on the many incongruities which observation of facts would reveal in this arrangement. I will only remark that the simple occurrence of the presence and volume of the said vesicle and of the allantois would give rise to not a few contradictions, or at least to very serious difficulties in its rigorous application.

Later anatomists remarked that in some animals it was the allantois alone that took part in the formation of the placenta, while in others, it, as well as the umbilical vesicle, had an important share. This fact, which I followed with some degree of attention in the *Cheiroptera* and some of the rodents, seemed to me of great value, and deserving of more careful consideration than to be allowed to remain in the field of simple anatomical observation; adding, then, my own investigations to those of others, I endeavored to ascertain what would result from separating the mammalia into *Allantoidea* and *Omphaloidea*, according as into the constitution and the vascularization of the placenta the allantois alone entered, or both the allantois and the umbilical vesicle. The following are the results, briefly stated, which were thus obtained. I have noticed only those groups in which in some species of the different genera the forms of the placenta are known. The table might serve for a natural zoölogical arrangement as well as for the phylogenetic order of the mammalia, not progressive in a direct series, but progressive, and in lines diverging from each other.

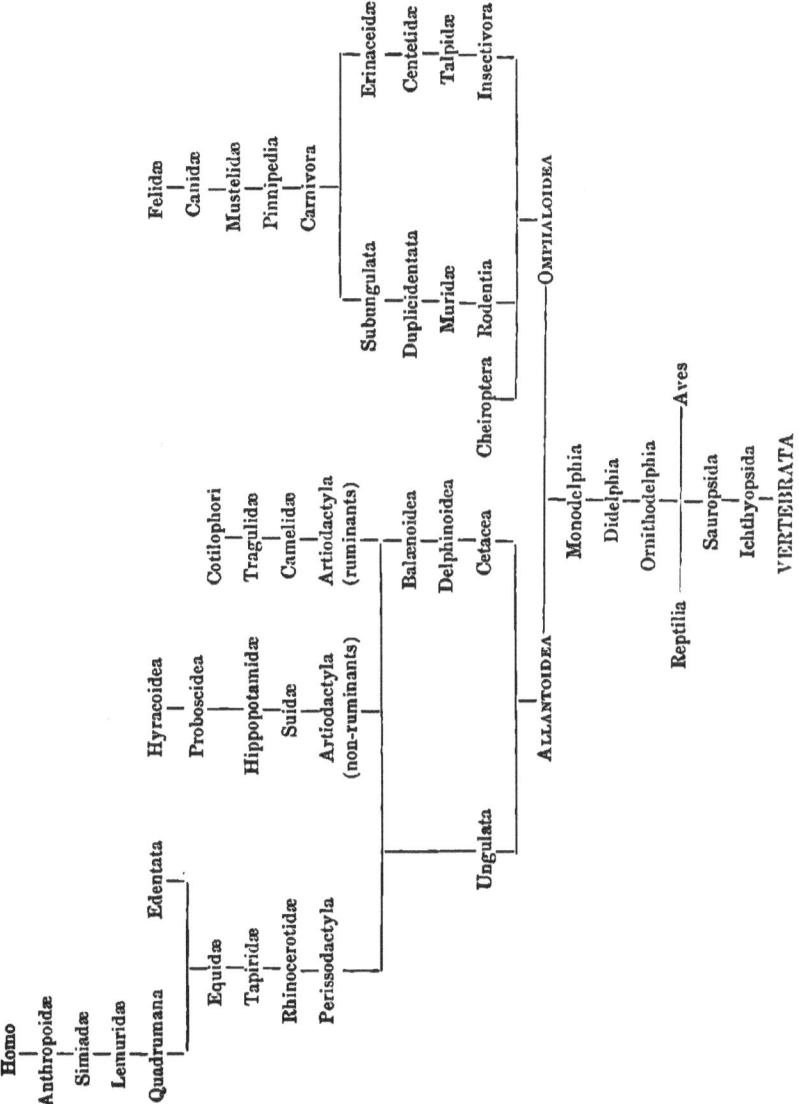

I will not now enter into even an outline of the arguments of various kinds with which I seek to support the basis of the general idea running through this attempt at a zoölogical arrangement which may reconcile embryology with the taxonomy and phylogeny of mammals. The inferences from it are many and important, but I shall confine myself to asking attention to a few only of the clearest actual results shown by comparison of the diverse forms of placenta as observed in the different animals which are referred to the groups indicated in the table. Especially will it be noticed that the various forms of placentæ ordinarily included under the generic name of diffused placentæ are observed only in the *Allantoidea*, and that there is not a single example of these forms among the *Omphaloidea*. In these latter there are not a few examples known of those forms of placenta which I have called cotyledonous with incomplete vascularization, and of this form there is not a single example among the *Allantoidea*.

In the *Canidæ* and *Felidæ*, the highest of the *Omphaloidea*, though the umbilical vesicle takes no direct part in the formation of the placenta, as is the case in *Cheiroptera* and *Rodentia*, it yet not only remains very voluminous to the close of gestation, but during that period increases in size with the development of the embryo, to indicate the ancient common origin of these animals with the *Ichthyopsida* and the *Sauropsida*. In the *Omphaloidea* the intimate structure is elevated and perfected in acquiring greater size, of which the single and discoidal placenta, though of villous structure, in the mole is a fine example, especially taken with the zonarial but perfected vascular placenta in *Canidæ* and *Felidæ*. In the *Allantoidea*, on the contrary, the placenta exhibits in all the groups the simplest form of diffused placenta, and is developed in structure by localization; and if we consider the ascending series in the genera of the different groups, we easily perceive the verification of a fact of some interest, namely, that the lowest animals in every group have the placenta of diffused form, and that in proportion as the animals in the same group rise, it always tends to become localized, until it is single and discoidal in the group of *Quadrumana*.

Modern zoölogists, having observed that in the *Lemuridæ* the placenta is diffused, have proposed to form, on this account only, a distinct order of these mammals, separating the *Lemuridæ* from the *Quadrumana*, — and anthropologists have made use of this same observation to combat the phylogenetic order taught by Haeckel, which, in the process of evolution, makes the lemurs precede the monkeys. In the order of *Quadrumana* is repeated the same fact noticed in all the groups of the *Allantoidea*, and the doubts of zoölogists and the observations of anthropologists lose, therefore, it appears to me, their importance. But carrying the investigations still farther it will not be difficult to discover another fact, and one of real importance to phylogeny, namely, that all living animals that have the simplest forms of placenta, whether *Allantoidea* or *Omphaloidea*, have their representative in the fossil animals found in the oldest strata of the earth; and, on the other hand, those which present the highest forms of placenta were the last to appear on the surface of the globe, — which furnishes a link that might not be suspected between embryology and paleontology.

Difficulty is met with in following the arrangement I have pointed out, when we reach the group of the *Edentata*, but, happily, we are not the only ones who strike on this rock. All the most distinguished zoölogists regard these animals as forming what they call a heterogeneous group, kept together because it is not known where to place them, some being deciduate and others non-deciduate. But even in this group of animals we see the fact repeated and maintained which we have observed in all the other groups of the *Allantoidea*: that some, as in the genus *Manis*, the Pangolins, have the placenta diffused; that in others it is perfected by localization, so far as to have the highest vascular structure with lacunose ectasia of the maternal vessels, as is seen in *Dasypus* and in *Tamandua*. It may be observed that all these animals appeared very late on the earth, and I venture to suggest, if, instead of repeating that the *Edentata* form a heterogeneous group, which really means nothing, we may not rather suppose that the progressive evolution of beings having arrived at the *pro-Quadrumana*, these diverged into two groups, in the one progressive, which from the *Quadrumana* arrived at the apes and man, in the other atavistic, tending towards the older forms of animals, in which the placenta exhibits an irregular and incomplete development.

INDEX OF AUTHORS.

	PAGE
Albinus, Bernhard S. (German anatomist and physician, 1696–1770)	79, 114
Aldrovandi (Bologna, 1621)	48
Andreini, Dr. R.	147
Arantius (Julius Cæsar, Bologna, 1530–1589)	79, 108
Aristotle (Greek philosopher, 384–322 B. C.)	33, 47, 48, 52
Aveling, Dr. J. H.	105, 106
Baer, Karl E. von (1792–1846)	7, 8, 29, 33, 36, 69, 75, 279, 289, 291
Barkow	12
Bartholin, Thomas (Copenhagen, 1616–1680)	79
Bassi, Professor (Turin)	174
Belluzzi, Dr.	117
Bernard, Claude (professor of physiology in Paris, 1813–1879)	55, 79, 80, 81
Berres	34
Bischoff, Theodore Ludwig (Munich, 1807)	9, 10, 13, 38, 41, 70, 96, 104, 123–125, 134, 187, 256
Blokham	124
Boivin, Madame (Brussels, 1834; 1773–1841)	74
Bojanus	108
Bruch, Prof. R. (Preface, page iv.)	277
Bruck, Prof. R. (Strasburg, 1860)	147, 155, 221, 222, 224
Burkhardt (Basle, 1834)	8, 10, 25
Burns (London)	109
Cavolini (Naples, 1787)	220
Cohnheim	152
Colin (Paris, 1856)	11, 27, 34, 35, 57
Colombo, Realdo (Venice, 1559)	65
Coste, Jean F. (French physician, 1741–1819)	125
Cowper, William (English anatomist, 1666–1709)	72

INDEX OF AUTHORS.

	PAGE
Creighton	273
Cuvier (French naturalist, 1773–1838)	76, 86, 234, 235, 237

Diocles 47
Duvernoy, George Louis (French anatomist and zoölogist, 1777–1855) 53

Edwards, Milne 274
Engelmann. Dr. George J. (St. Louis, 1875) 106
Eschricht (Copenhagen, 1837) . 8, 12, 24, 53, 74, 96, 125, 202, 237
Everard (Middelburg, 1661) 79

Fabricius, Girolamo of Acquapendente (Italian anatomist. Padua, 1537–1619) . . . 29, 32, 33, 48, 51, 52, 57, 66, 70, 72, 80, 178
Fallopius (Italian anatomist. 1523–1562) 108
Farre, Arthur (Cyclopedia of Anat., London, 1858) 114–116, 126, 259
Florinsky (St. Petersburg, 1863) 163

Galen, Claudius (Rome, 131–201) 47, 53, 71
Gamgee, Dr. Arthur (Edinburgh, 1864) 54, 55
Graaf, Regnier van (Leyden, 1641–1678) . . . 31, 63, 78
Gurlt, Professor (professor of surgery in Berlin, 1860–1868) 11, 13, 84

Haller, Albert (Berne, 1708–1777) 29, 32, 53, 68, 77, 93, 109, 113–273
Harting 273
Harvey, William (London, 1578–1658) . . 52, 54, 57, 66, 78
Hildebrand (Brunswick, 1832) 8
Hippocrates (Athens, 460–377 B.C.) 47, 78
Hirlt (Milan, 1858) 15, 16, 73
Hobokenius (Utrecht, 1672) 48, 50, 60, 109, 113
Home, Sir Everard (1756–1832) 290
Hunter, William (London, 1718–1783) . 66, 105, 123, 124, 256
Huxley, Prof. Thomas H. (London, 1825) . 219, 223, 279, 290

Kirvisch 251
Knok (London, 1840) 124
Kölliker, Rudolf Albert (professor of anatomy in Würzburg; Leipzig, 1876) . . 177, 181, 210, 224, 225, 227, 252, 256, 258–260
Kolk, Schroeder van der 252

Langhans (1870) 256
Lauth (Geneva, 1657) 78, 79
Leydig (Frankfurt, 1857) . . . 12, 22, 40, 85, 160, 221
Lieberkuhn 12

INDEX OF AUTHORS.

PAGE

Malphigi, Marcello (Bologna, 1628–1694) 6–8, 12, 13, 20, 48, 49, 52, 62, 68, 159, 177
Monro, Alexander (Edinburgh, 1696–1767) 72
Morin 55
Müller (Paris, 1851) 27, 30, 33, 36, 75, 80, 86, 109
Müller, Johannes (Vienna, 1801–1858) . . . 73, 221, 284
Myddleton 12

Needham (London, 1767) . . 30, 33, 48, 51, 52, 67, 80, 81, 113
Noortwyck (Leyden, 1743) 65

Owen, Richard (London, 1804) . 177, 181, 224, 225, 227–229, 290

Panizza (Milan, 1866) 20, 49, 50, 62, 63, 72
Piano, Dr. G. Pietro (Bologna) 195
Pouchet, Felix A. (Paris, 1800–1872) 112
Prévost 55

Ramsbotham (London, 1834) 66
Radford (Manchester, 1832) 66
Reid 125
Rivolta, Professor (Turin) 174
Robin, Charles (professor of anatomy in Paris, 1858) 112, 113, 118, 250, 254
Rolleston 290
Rossi (Bologna) 214
Rouhault (member of Royal Academy of Sciences, 1714) . . 109
Ruini (Bologna, 1596) 28, 33, 35, 42

Schiff, Professor (Paris, 1866) 80
Schlossberger (Tübingen, 1855) 54
Seiler 104, 105
Severi, Dr. (1832) 59
Severin, Marcus Aurelius (Nuremberg, 1645) 48, 80
Sharpey, William (professor of anatomy, London) 10, 13, 17, 34, 69, 70, 75, 105
Sinèty, De (Paris, 1876) 259
Snape (London, 1606) 31
Spiegelberg (Freiburg, 1864) . . . 12, 13, 48, 49, 55, 59, 187
Stuart 113

Trinchese, Prof. Salvatore 222
Turner, Prof. William (Edinburgh, 1876) 181, 187, 202, 210, 228, 230, 232, 236, 237, 245, 273, 290

	PAGE
Velpeau, M. Alfred (Brussels, 1795–1867)	66, 69
Vesalius, Andreas (1514–1564)	53
Vieussens, Raymond (1641–1720)	53
Vierordt (Milan, 1857)	82, 103, 104, 110, 115
Virchow, Rudolf (professor of pathological anatomy in Berlin, 1821–)	165, 251
Waldeyer (professor of pathological anatomy in Breslau)	181, 187, 247
Weber brothers	7, 8
Weber, E. H. (Leipzig, 1827–1867)	8, 10, 11, 16, 34, 70, 98, 104, 106, 124, 125, 184, 257, 288
Wepfer	31
Wharton, Thomas (Amsterdam, 1659; 1610–1673)	67, 68, 78, 80
Winkler	255
Williams, Dr. John (1875)	106

INDEX OF SUBJECTS.

Absence of the placenta and of the cotyledons in the gravid uterus of certain animals maintained by the ancients, 17.
Acanthus vulgaris, 282.
Achoria, mammalia, 225.
Acotyledonous placenta: simple, 277-280; villous and diffused, 281; complicated villous, 282; localized villous, 282.
Albumen, one of the constituent elements of the cotyledonal fluid, 54.
Albuminoid substances, 82.
Alkaline albuminates, 55.
Allantoidea, the placenta exhibits in all the groups the simplest form of diffused placenta, and is developed in structure by localization, 292-296.
Allantois of the ruminants, 13; in kangaroo, 228; increase of, 291, 292.
Alveolar surface, 51.
Alveoli of the cotyledons, 50.
Amorphous matter of the grayish tissue of the uterine surface, 112.
Amylaceous matter, 80.
Analysis of the anatomical knowledge of the ancients relative to diffused placenta, 27-48; of single placenta, 34-40; of multiple placenta, 48-56; of the fluid secreted by the uterine cotyledons, 55.
Anatomical structure of the placenta in the cow, the sheep, and the hind, 151-160; truths taught by the ancients concerning the uterine mucous membrane, 32-38; structure of the villi, 116-122.
Anatomists' opinions upon the utricular glands, 6-11; opinions upon the uterine mucous membrane, 27-40.
Anthropoid apes, 287.
Anthropoidæ, 293.
Anthropologists, observations of, 295.
Apertures in the septa, 251.
Appendix, 147-174
Arteries, 251.
Artiodactyla, 293.
Aves, 293.

Baer, von, classification of mammals by, 291; doctrine regarding placental development, 279.

Balænoida, 293.
Balænoptera Sibbaldii, 281.
Batrachians, intestinal glands of, 12.
Bos taurus, 282.
Bradypus didactylus and tridactylus, 286.
Branching utricular glands, 10-15.
Bursa, chorial, 197.

Calices of the utricular glands, 62; of the new glandular organ, 153.
Camelidæ, 293.
Camelopardalis giraffa, manner of forming placenta in, 282.
Camelus dromedarius, 281.
Canaliculi, glandular structure of, 8-20.
Canaliculus, 20.
Cancerous cells, so-called, 112.
Canidæ, 293.
Canis vulpes and domesticus, 285.
Capillaries, net-work of, 253; of the villi, 261.
Capillary blood-vessels, 16.
Capra hircus, 287.
Carcharias, rudimental placenta of, 219.
Carnivora, observations upon the glandular organ of the, 140; intimate structure of the placenta in, 188.
Caruncles of the chorion separate easily from the uterine cotyledons in the development of the fœtus, 48.
Caseine, one of the constituent elements of the cotyledonal fluid, 54.
Cat, changes in the uterine mucous membrane of, 158-164; formation of new glandular organ in the, 163.
Catamenial decidua, established during menstruation, and highly developed during conception, 105; the product of materials elaborated by the utricular glands, 134.
Cavia cobaya, placental development of, 167; transverse section of placenta of (Plate IV., Fig. 1), 198.
Cells, metamorphoses of serotinal, 118; of embryonic, 82; of glandular, 80.
Centetes ecaudatus, 286.
Centetidæ, 293.
Cercopithecus sabeus, 287.
Cervus dama, capreolus, elaphus, axis, porcinus, mexicanus, 283.

INDEX OF SUBJECTS.

Cetacea, 293.
Cetaceans, uterine glands of, 10.
Changes in the ovum in the rat during gestation, 193–198; in the uterine mucous membrane of the Cavia cobaya during gestation, 198–205.
Cheirogaleus, 281.
Cheiroptera, 292.
Chemical composition of the cotyledonal fluid, 54.
Cholæpus Hoffmanni, 286.
Choriata, mammalia, 225.
Chorion, villi of, 23–35.
Chyliferous vessels, 79.
Classification of the mammalia, 272–296; table of, 293.
Conclusions upon the utricular glands and mucous membrane of the uterus, 130–136; upon the glandular organ of new formation or maternal portion of the placenta, 137–145; relative to the development of the decidua and placenta in animals and in the human species, 206–209; relative to the destructive and neo-formative processes, 265–271.
Cotilophori, 293.
Cotyledonous placentæ of incomplete vascularization, 151–157; with complete vascularization, 285; with complete vascularization and ectasia of vessels, 286.
Cotyledons, maternal, 13; rudimentary, 18; diameter of uterine, 21; differences in volume during pregnancy, 48; development of uterine, 58; anatomical structure and functions of, in the cow and the hind, the, 57.
Cow, rudimentary cotyledons of, 13–21; formation of placenta in, 150–153.
Cryptal form of the new glandular organ, 281.
Crypts, simple glandular, 10–18; placental in cat, 187–190.
Cyclothurus didactyla, 287.
Cynocephalus sphinx, lacunose ectasia in the maternal vessels of, 287.

Dasyprocta aguti, 285.
Dasypus gymnurus, novemcinctus, and sexcinctus, 286.
Decidua, catamenial, 105, 134; human, 25; reflected, 107; serotina, 111–116; uterine and reflected, fusion of the two, 107; placentalis sub-chorialis and sensu-strictiori, 255; early researches upon, 112.
Deciduæ reflected, serotina and vera, structure and function of, in the Cavia cobaya, 196–202.
Defenders of the doctrine of the direct communication of the blood of the mother with that of the fœtus, 11–82.
Delphinoidea, 293.
Delphinus phocæna, 281.
Demonstration of the fact that there exist two species of uterine glands in some animals, 18.
Destruction of the old uterine mucous membrane, 193.
Destructive processes carried on during gestation, 196.
Development of the placenta, 93–101; of the glandular organ of neo-formation, 138; of the utricular glands, 14–26.
Diagrams representing vertical sections of the uterus and placental development of animals and of the human species. (See Atlas.)
Diameter of transverse canal and of the utricular glands, 14–21.
Diaphanous fibrous tissue, 119.
Didelphia, 280, 293.
Differences in gravid and non-gravid uterus of torpedo, 220; in the diameter of the utricular glands, 21.
Dilatation of the vascular loop during gestation, 170; of utricular glands, 11–23.
Discoidal form of the placenta, 282–286.
Distinctions between the human placenta and that of animals, 121–129; among mammals, 225.
Dog, typical form of the glandular follicles preserved in the placenta of, 93; section of the placenta of dog at term, 190.
Dolphin, uterine glands of, 9.
Duplicidentata, 293.

Ectasia of chorial villi, 176; in maternal vessels, 268; in utero-placental vessels, 287.
Ectasis of maternal vessels, 202.
Edentata, some called deciduates, some non-deciduates, 289, 293.
Elephas indicus, example of complicated villous placenta, 281.
Embryo, development of, 82; nutrition of, 218.
Embryonic cells, 82.
Endosmose, fœtal nutrition carried on by, 137.
Envelopes, fœtal, 28; of the parietal cells, 251.
Epithelial buds of the villi, 261; cells, 81; layer, 14, 19.
Epithelium, internal, 14; of old mucous membrane, 196; of the villi, 261; of maternal secretory vessel, 241; of the chorial villi, 176.
Equus caballus and asinus, 281; example of follicular placentæ, 285.
Erinaceadæ, 15.
Erinaceidæ, 293.
Erinaceus europeus, example of discoidal form of placenta, 285.
Exosmose, fœtal nutrition carried on by means of endosmose and, 137.
Experiments upon the uterine mucous membrane, 16–20.

INDEX OF SUBJECTS. 303

Expulsion of glandular organ in delivery, 142.
Exterior membrane surrounding the villi, 116.
External form of placentæ, 287.

Factors upon which the essential differences in the structure of the two villi in the various forms of placenta in the mammalia depend, 239.
Fallopian tubes, 104.
Fecundation, 92.
Felidæ, 293.
Felis catus domesticus, of zonarial form, 285.
Fibrine, one of the constituent elements of the cotyledonal fluid, 54.
Filaments, lymphatic, 74.
Fœtus, nutrition of, 137, 172.
Folds of the mucous membrane, 92.
Follicles, simple, glandular, 2, 10, 18, 80; type of these, 42.
Function of cotyledons in ruminants, 52; in animals with single placenta, 65.
Fusiform cells, 116.

Genesis of new vessels by vascular transformation, 128.
Gestation, changes occurring in the mucous membrane during the period of, 15-20, 193.
Glands, utricular, 9; two species of uterine, 16; increase in volume during gestation, 8-12; secretion of, 44.
Glandular organ of new formation, 27-46; cells of, 80.
Glycogenesis, 80.
Glycogenic cells, formation of, 56.
Guinea-pig, development of the placenta in, 167; description of the fully formed placenta in, 198-202.

Halichærus grypus, 283.
Halicore dugong, 281.
Hapale jacchus, 287.
Hapalemur, 281.
Hare, development of the placenta in, 166.
Hepatic placental organ, 80.
Hind, development of the cotyledons, 155.
Hippopotamidæ, 293.
Hippopotamus amphibius, 282.
Histogenetic process by which glandular follicles are formed, 140.
Histologists, opinions of, relative to the utricular glands, 6-16; the glandular organ, 26-38; in animals with multiple placenta, 47-57; in animals with single placenta, 65-81; upon the human placenta, 102-112, 177-181.
Homo, 293.
Human decidua, a product of exudation, 25.
Human placenta, structure of, 102-129; peculiarities of, 121; lacunar circulation of, 127.
Hyomoschus aquaticus, 281.
Hyperplasia and hypertrophy of sub-connective tissue. 117-121.
Hyracoidea, 293.
Hyrax capensis, cryptal form of placenta in, 283.

Ichthyopsida, 293.
Implacentalia, applied to the marsupialia; monotremia, cetaceans, 225.
Insectivora, 234, 293.
Introflexion of the walls of the maternal vessels upon the villi, 257; of the epithelial layer, 130.

Lacunæ, formation of, 257; upon the internal portion of the uterine surface of the placenta, 122; placental, 145.
Lacunar circulation of the human placenta, 127.
Lacunose ectasia in the maternal vessels, 287; circulation, 145.
Lamellæ, uterine, formed by the elevation of new tissue, 161, 162; representation of, Plate X., Fig. 1 (g, g, g).
Lamina of the sub-chorial tissue, 251.
Lemuridæ, 293.
Lemur rufipes, villous placenta, 281.
Lepilemur, villous placenta, 281.
Lepus caniculus and timidus, 285.
Lesion, traumatic, of uterus, 142; pathological, of placenta, 121.
Lumen of utero-placental vessels, 184.
Lutra vulgaris, 283.

Macacus nemestrinus and cynomolgus, 287.
Macropus major, 227, 280.
Mammalia, different periods of nutrition in, 82; one law governing the formative process of the placenta and the nutrition of the fœtus in all the, 217; embryonic phases of, 121; classification into Allantoidea and Omphaloidea of all the, 293; factors upon which depend the diverse forms of placenta in the, 239; two fundamental parts, one vascular and absorbent, the other maternal and secretory, in the placenta of all, 224; nutrition of the fœtus during intrauterine life, 137; classification into deciduates and non-deciduates, 279.
Mammals, species of glands in, differing in structure, volume, and office, in all, 24, 131; single anatomical type of the placenta in all, 184; multiparous, 182; with diffused placenta, 212; classification of, 291, 293.
Mammifera, uterine glands in, 6-27; implacentalia and placentalia, 224-227.
Manis, 296.
Mare, utricular glands and uterine mucous membrane of, 6-38; development

304 INDEX OF SUBJECTS.

of placenta, 97; vertical sections of gravid and non-gravid uterus, 36; changes over the whole uterine surface during gestation, 42-46.

Marsupialia, classed by Owen with implacentalia, 225; simplest form of placenta joined with the simplest but relatively higher form of diffused placenta found in the, 230-237.

Membrane, existence or non-existence of the uterine mucous, 84; changes in the same, 106; functions of, 135.

Memoir, placental development in all mammals, including the human species, 6-146.

Metamorphoses of the cells of the serotina, 118.

Modifications of the new glandular organ in certain mammals, 140.

Mole, uterine cotyledons of the, 154.

Monachus albiventer, 283.

Monkey, placental development in, 174.

Monodelphia, 293.

Monodon monoceros, 281.

Monograph, unity of type in placental development, 176-271.

Monotremia, 225.

Muridæ, 293.

Mus decumanus, 285; transverse section of a gravid uterus of, 193.

Mus musculus, 285; segment of the gravid uterus of, 195.

Mustela foina, martes, and vulgaris, 283.

Mustelidæ, 293.

Mustelus lævis, description of uterine horn, 219-222.

Mycetes ursinus, 287.

Neoplasm, cellulo-vascular, 193-195.

Noctilio leporinus, 286.

Nutrition, fœtal, 54, 82; during intra-uterine life, 137.

Observations of the ancients upon the utricular glands, 6-13; upon the villi of the chorion, 27-33; upon the uterine cotyledons, 47-52; upon the structure and formation of the placenta, 65-85.

Omphaloidea, 292-294.

Orca gladiator, 281.

Ornithodelphia, 280, 293.

Ornithorhynchus paradoxus, 280.

Orycteropus, 288.

Oviparous vertebrates, 219.

Ovis aries, 292.

Ovum, nutrition of the, 82; changes during gestation, 193; development of the, 163.

Pachyderms, uterine glands of, 10; fœtal and maternal placenta of, 29-31.

Pangolinus, 281.

Papillæ, placental, 27-34.

Pedicle of the cotyledon, 60-64.

Peduncle of the placenta, 200-204.

Perissodactyla, 293.

Phoca bicolor and vitulina, 283.

Phyllostoma hastatum, 286.

Phylogeny, importance of the study of the structure of the placenta relative to zoölogical, 290-296.

Pinnipedia, 293.

Placenta, multiple, 47-64; single, 65-101; human, 102-129; conclusions relative to the maternal portion of, 137-145; formation of the glandular portion of, 147-174; histological unity of the anatomical type of, 176-218; unity of the physiological processes and office of the, 218-271; microscopical studies showing the unity of type of the, 277-296.

Placentæ, simple acotyledonous, 280; villous and diffused acotyledonous, 280; complicated villous acotyledonous, 281; localized villous acotyledonous, 282-284; cotyledonous, 284; cotyledonous, of incomplete vascularization, 285; cotyledonous, of complete vascularization 285.

Plagiostomi, acotyledonous, 279; cotyledonous, 279.

Proboscidea, 293.

Processes, destructive, 193; neo-formative, 265.

Propithecus verrauxii, 281.

Pteroplatea altavela, villi of mucous membrane of the, 221.

Pteropus medius, 286.

Quadrumana, classification of, 293-294.

Rabbit, uterine glands, 22-25; development of the placenta of, 90-93; changes in the uterine mucous membrane, 88-96.

Raja torpedo, embryo of, 220.

Rangifer tarandus, 283.

Reptilia, 293.

Rhinocerotidæ, 293.

Rodentia, 293.

Ruminants, difference of opinion as to the cotyledons of, 51-56; development of maternal cotyledons of, 58-64.

Sauropsida, 293, 294.

Scymnus lichia, 222.

Selachii, villous neo-formation in the mucous membrane of the gravid uterus of, 277; viviparous, 219.

Semnopithecus nasica and mitratus, 287.

Serotina, differences of opinion as to the character and function of the decidua, 108-114.

Sheep, rudimentary cotyledons of, 154.

Simiæ, 287-291.

Simiadæ, 293.

Sinuses, placental, 122.

Solipeds, lacteal fluid of, 281.

Squalidæ, placenta of, 76.

Subungulata, 293.
Suidæ, 293.
Summary, including anatomical facts as to the most remarkable differences seen in the placenta of the mammifera, with their application to taxonomy and phylogony, 272-296.
Sus scrofa, 281.

Talpa europæa, 284.
Talpidæ, 293.
Tamandua, 296.
Tamandua tetradactyla, 286.
Tapiridæ, 293.
Taxonomy, reconciliation of the embryology of mammals with, 294.
Trabecæ, cellulo-vascular, 197.
Tragulidæ, 293.
Tragulus Stanleyanus and meminna, 281.
Troglodytes niger, 287.
Tunnel, peripheral part of cotyledonal, 203.

Ungulata, 293.
Uterus, utricular glands and uterine mucous membrane of, 6-26; formation of new glandular organ in the gravid, 27-46; conclusions relative to the utricular glands and mucous membrane of, 130-136.

Veins, circulation of maternal blood in, 124; utero-placental, in human species, 144.
Vertebrata, 293.
Vertebrates, oviparous, 219; mammiferous, 224.
Vespertilio murinus and noctula, 285.
Vessels, injection of fœtal, 190; utero-arterial and venous, 191; placental, 44.
Villi, chorial, 10-30; structure of placental, 113-121.

Woman, development of the placenta in, 169-173.

www.ingramcontent.com/pod-product-compliance
Lightning Source LLC
Chambersburg PA
CBHW031903220426
43663CB00006B/741